T0236842

Lecture Notes in Computer Science 10543

Commenced Publication in 1973
Founding and Former Series Editors:
Gerhard Goos, Juris Hartmanis, and Jan van Leeuwen

Editorial Board

David Hutchison
  Lancaster University, Lancaster, UK
Takeo Kanade
  Carnegie Mellon University, Pittsburgh, PA, USA
Josef Kittler
  University of Surrey, Guildford, UK
Jon M. Kleinberg
  Cornell University, Ithaca, NY, USA
Friedemann Mattern
  ETH Zurich, Zurich, Switzerland
John C. Mitchell
  Stanford University, Stanford, CA, USA
Moni Naor
  Weizmann Institute of Science, Rehovot, Israel
C. Pandu Rangan
  Indian Institute of Technology, Madras, India
Bernhard Steffen
  TU Dortmund University, Dortmund, Germany
Demetri Terzopoulos
  University of California, Los Angeles, CA, USA
Doug Tygar
  University of California, Berkeley, CA, USA
Gerhard Weikum
  Max Planck Institute for Informatics, Saarbrücken, Germany

More information about this series at http://www.springer.com/series/7410

Pooya Farshim · Emil Simion (Eds.)

Innovative Security Solutions for Information Technology and Communications

10th International Conference, SecITC 2017
Bucharest, Romania, June 8–9, 2017
Revised Selected Papers

 Springer

Editors
Pooya Farshim
École Normale Supérieure
Paris
France

Emil Simion
Polytechnic University of Bucharest
Bucharest
Romania

ISSN 0302-9743   ISSN 1611-3349   (electronic)
Lecture Notes in Computer Science
ISBN 978-3-319-69283-8   ISBN 978-3-319-69284-5   (eBook)
https://doi.org/10.1007/978-3-319-69284-5

Library of Congress Control Number: 2017956772

LNCS Sublibrary: SL4 – Security and Cryptology

© Springer International Publishing AG 2017
This work is subject to copyright. All rights are reserved by the Publisher, whether the whole or part of the material is concerned, specifically the rights of translation, reprinting, reuse of illustrations, recitation, broadcasting, reproduction on microfilms or in any other physical way, and transmission or information storage and retrieval, electronic adaptation, computer software, or by similar or dissimilar methodology now known or hereafter developed.
The use of general descriptive names, registered names, trademarks, service marks, etc. in this publication does not imply, even in the absence of a specific statement, that such names are exempt from the relevant protective laws and regulations and therefore free for general use.
The publisher, the authors and the editors are safe to assume that the advice and information in this book are believed to be true and accurate at the date of publication. Neither the publisher nor the authors or the editors give a warranty, express or implied, with respect to the material contained herein or for any errors or omissions that may have been made. The publisher remains neutral with regard to jurisdictional claims in published maps and institutional affiliations.

Printed on acid-free paper

This Springer imprint is published by Springer Nature
The registered company is Springer International Publishing AG
The registered company address is: Gewerbestrasse 11, 6330 Cham, Switzerland

Preface

This volume contains the papers presented at SecITC 2017, the 10th International Conference on Security for Information Technology and Communications (www. secitc.eu), held during June 8–9, 2017, in Bucharest. There were 22 submissions and each submitted paper was reviewed by at least three Program Committee members. The committee decided to accept seven papers (one paper was withdrawn by the authors, after the conference, from the LNCS volume) as well as a further seven invited speakers. For ten years SecITC has been bringing together computer security researchers, cryptographers, industry representatives, and graduate students. The conference focuses on research on any aspect of security and cryptography. The papers present advances in the theory, design, implementation, analysis, verification, or evaluation of secure systems and algorithms. One of SecITC's primary goals is to bring together researchers belonging to different communities and provide a forum that facilitates the informal exchanges necessary for the emergence of new scientific collaborations. We would like to acknowledge the work of the Program Committee, whose great efforts provided a proper framework for the selection of the papers. The conference was organized by Advanced Technologies Institute, Bucharest University of Economic Studies and Military Technical Academy.

July 2017

Pooya Farshim
Emil Simion

Foreword

It is a priviledge for me to write the foreword to the proceedings to this 10th anniversary of the conference. Indeed, SECITC 2017 is the 10th edition of the International Conference on Information Technology and Communication Security held in Bucharest, Romania every year.

Throughout the years, SECITC has become a truely competitive publication venus with an acceptance rate of 1/3, an Program Committee of 50 experts from 20 countries and a long series of distinguished invited speakers. Since three years the conference proceedings are published in Springer's Lecture Notes in Computer Science, and articles published in SECITC are indexed in most science databases.

The conference is unique in that it serves as an exchange forum between confirmed researchers and students entering the field as well as industry players.

I would like to particularly thank the PC chairs Pooya Farshim and Emil Simion for an outstanding paper selection process conducted electronically. In response to the call for papers the Program Committee got 22 submissions of which seven were chosen. To those the PC added seven invited keynote lectures by Sylvain Guilley, Konstantinos Markantonakis, Claudio Orlandy, Peter Ryan, Ferucio-Laurentiu Tiplea, Damien Vergnaud, and myself.

I also warmly thank the conference's Organization Committee and Technical Support Team Mihai Doinea, Cristian Ciurea, Luciana Morogan, Andrei-George Oprina, Marius Popa, Mihai Pura, Mihai Togan, and Marian Haiducu for their precious contribution to the success of the event and for their dedication to the community.

I am certain that in the coming years SECITC will continue to grow and expand into a major cryptography and information security venue making Bucharest a traditional summertime scientific meeting habit to the IT security research community.

August 2017   David Naccache

Organization

Program Committee

Elena Andreeva	COSIC, KU Leuven, Belgium
Ludovic Apvrille	Telecom ParisTech, France
Gildas Avoine	INSA Rennes, France; UCL, Belgium
Manuel Barbosa	HASLab - INESC TEC and FCUP
Ion Bica	Military Technical Academy, Romania
Catalin Boja	Bucharest Academy of Economic Studies, Romania
Sanjit Chatterjee	Indian Institute of Science, India
Liqun Chen	University of Surrey, UK
Christophe Clavier	Université de Limoges, France
Paolo D'Arco	University of Salerno, Italy
Joan Daemen	STMicroelectronics and Radboud University in Nijmegen, The Netherlands
Roberto De Prisco	University of Salerno, Italy
Eric Diehl	Sony Pictures, USA
Itai Dinur	Ben-Gurion University, Israel
Stefan Dziembowski	University of Warsaw, Poland
Pooya Farshim	ENS, France
Bao Feng	Huawei, China
Eric Freyssinet	LORIA, France
Nicolas Gama	University of Versailles, France
Helena Handschuh	COSIC, KU Leuven, Belgium
Shoichi Hirose	University of Fukui, Japan
Xinyi Huang	Fujian Normal University, China
Miroslaw Kutylowski	Wroclaw University of Technology, Poland
Jean-Louis Lanet	Inria-RBA, France
Giovanni Livraga	Università degli Studi di Milano, Italy
Konstantinos Markantonakis	ISG-Smart Card Centre, Founded by Vodafone, G&D and the Information Security Group of Royal Holloway, University of London, UK
Florian Mendel	TU Graz, Austria
Bart Mennink	Digital Security Group, Radboud University, Nijmegen, The Netherlands
Kazuhiko Minematsu	NEC Corporation, Japan
David Naccache	ENS, France
Rene Peralta	NIST, USA
Bart Preneel	KU Leuven COSIC and iMinds, Belgium
Reza Reyhanitabar	NEC Laboratories Europe, Germany
P.Y.A. Ryan	University of Luxembourg, Luxembourg

Damien Sauveron	XLIM, UMR University of Limoges/CNRS 7252, France
Emil Simion	University Politehnica of Bucharest, Romania
Agusti Solanas	Smart Health Research Group, Rovira i Virgili University, Spain
Rainer Steinwandt	Florida Atlantic University, USA
Willy Susilo	University of Wollongong, Australia
Ferucio Laurentiu Tiplea	Alexandru Ioan Cuza University of Iasi, Romania
Mihai Togan	Military Technical Academy, Romania
Cristian Toma	Bucharest Academy of Economic Studies, Romania
Denis Trcek	University of Ljubljana, Slovenia
Michael Tunstall	Cryptography Research, USA
Victor Valeriu	Military Technical Academy, Romania
Serge Vaudenay	EPFL, Switzerland
Ingrid Verbauwhede	ESAT - COSIC, Belgium
Guilin Wang	Huawei International Pte. Ltd., China
Qianhong Wu	Beihang University, China
Lei Zhang	East China Normal University, China

Additional Reviewers

Balasch, Josep
Balli, Fatih
Bogos, Sonia
Chen, Siyuan
Li, Jiangtao
Li, Letitia
Li, Yanan
Lugou, Florian
Maimut, Diana
Slowik, Marcin
Unterluggauer, Thomas
Werner, Mario
Wszola, Marta
Zhang, Wentao

Contents

Faster Zero-Knowledge Protocols and Applications
(Invited Talk Abstract)

Claudio Orlandi[✉]

Aarhus University, Aarhus, Denmark
orlandi@cs.au.dk

Abstract. *Zero-knowledge (ZK) protocols* are one of the cornerstones of modern cryptography. In a nutshell, a ZK protocol allows a prover P (with a secret input x) to persuade a verifier V that $f(x) = 1$ for some public function f, without disclosing to V any other information about x. In this talk I will present two recent ZK protocols, known as ZKGC [JKO13,FNO15] and ZKBoo [GMO16]. These are the first ZK protocols that allow to prove interesting, non-algebraic statements (such as "*I know x such that SHA-256$(x) = y$*" for a public y), in the order of tens of milliseconds on a standard computer. As ZK protocols are ubiquitous in cryptography, this line of research has already enabled many interesting applications. In particular, I will show how ZKBoo allows to construct post-quantum signature schemes using symmetric-key primitives [CDG+17] only.

1 Talk Summary

This talk contains a high-level overview of a recent line of research that deals with the design of efficient zero-knowledge (ZK) protocols for arbitrary languages and with their applications. The talk, and therefore this document, contains no previously unpublished research results.

1.1 Introduction to Zero-Knowledge Protocols

Zero-knowledge (ZK) protocols are one of the cornerstone of modern cryptography and were introduced by Goldwasser, Micali and Rackoff [GMR85,GMR89] in the mid-80s. As many other notions in modern cryptography (such as public-key encryption, secure multiparty computation or homomorphic encryption) ZK protocols allow to perform a counter-intuitive and seemingly impossible task.

A ZK protocol is a protocol between two parties, usually referred to as the prover P and the verifier V. For the sake of simplicity the goal of ZK protocols is here defined in a way which is different from the standard literature: we have a prover P that knows some secret x which satisfies some public and efficiently computable predicate f i.e., the prover "knows" a value x such that $f(x) = 1$ (we will return on what it means for a computer program to "know" something

© Springer International Publishing AG 2017
P. Farshim and E. Simion (Eds.): SecITC 2017, LNCS 10543, pp. 1–11, 2017.
https://doi.org/10.1007/978-3-319-69284-5_1

later on). As the name suggests, the verifier V is interested in verifying that the prover really knows this secret. However this should happen in such a way that the verifier *does not learn any information about the secret x*.

An example of a commonly used protocol where the verifier learns *a lot* about the secret x we consider the common password-based authentication mechanisms that is nowadays used on most websites. In this case the user plays the role of the prover and the server the role of the verifier. The user claims to know some password x and the server stores some hash of the password e.g., $y = h(x)$ which defines the predicate $f(x)$. In particular in this case $f(x) = 1$ iff $h(x) = y$.[1]

The current implementation of password-based authentication is typically the following: to prove that the user knows the password x, the prover sends the secret x to the server that can in turn verify that the password matches the hashed value. Clearly, this leaks much more information than intended! From one hand we tell users to keep their password secrets, and from the other hand we instruct them to send their secret to another party every time they want to prove their identity! This is not without unwanted consequences, and is exploited by attackers via (increasingly common) phishing attacks, in which a user is fooled into interacting with an adversarially controlled server. Therefore, as the user enters their password believing they are trying to login on a legitimate server, the adversary learns the user's password.

The main property of a ZK protocol is to avoid the above problem: a ZK protocol allows P (with secret input x) to persuade V that $f(x) = 1$ in such a way that V does not learn any other information about x. In a nutshell, a ZK protocol is a (potentially interactive) protocol which should satisfy the following properties:

Completeness: If P knows x s.t., $f(x) = 1$ and both P and V follow the protocol instructions then V will output "accept".

Proof-of-Knowledge: If P does not know a value x such that $f(x) = 1$, then V will output "reject" *even if P does not follow the protocol instructions*.

Zero-Knowledge: V learns only that $f(x) = 1$ (and nothing else about x) by interacting with P, *even if V does not follow the protocol instructions*.

Some comments about these properties[2]: a weaker version of the proof-of-knowledge (PoK) property is sometimes used, called *soundness*: A ZK protocol satisfies soundness if P cannot make V accept in the case that *there exist no* x such that $f(x) = 1$. Unfortunately this requirement is typically too weak for cryptographic applications. As an example, in the password-based authentication considered before it would not be enough for the prover to demonstrate that a password matching the hash exists (which is trivially true), but that the prover "knows" that password. The fact that a prover (e.g., a computer program) "knows" a piece of information can be formalized by requiring that if P makes V accept, then it is possible to "extract" the secret from P (possibly using techniques such as rewinding).

---

[1] Hashed password should always be "salted" but we ignore this here to keep the notation simpler.

[2] For a formal treatment of the definition of ZK protocols see e.g., [Gol01, Gol04].

1.2 An Example: Schnorr Protocol and the Fiat-Shamir Heuristic

One of the most popular ZK protocols is perhaps Schnorr protocol [Sch89], which allows to prove knowledge of discrete logarithms in a very efficient way. Given a cyclic group G of prime order q generated by g, Schnorr protocol allows a prover with secret x to persuade a verifier with input h that $h = g^x$. The protocol is so simple that can be described here, and will also allow to exemplify some of the concepts introduced so far:

1. P chooses a random value $r \leftarrow \mathbb{Z}_q$, computes $a = g^r$ and sends a to V;
2. V chooses a random bit e and sends it to P;
3. P computes $z = xe + r \mod q$ and sends it to V;
4. V outputs accepts iff $h^e a = g^z$.

It is easy to see that the protocol is *complete* since $g^z = g^{xe+r} = (g^x)^e g^r = h^e a$. One can also easily see that, after P has "committed" to a, P can only reply to both $e = 0$ and $e = 1$ if P "knows" x: if V chooses $e = 0$ then P must send $z_0 = r$ to make V accept, and if V chooses $e = 1$ then P must send $z_1 = x + r$ to make V accept. Thus, given accepting (z_0, z_1) it is possible to extract $x = z_1 - z_0 \mod q$. Now, since P can only reply to both challenges if P knows x, it follows that *if P does not know x*, then P can make V accept for at most one challenge e or, in other words, if P does not know x then V will output reject with probability at least $1/2$. This is clearly not good enough (a cheating prover has a significant chance of making V accept) but the probability can be reduced to 2^{-s} by repeating the protocol s times. The property here described is typically referred to as *special soundness* and, as in can be seen, it is tightly related to the *proof-of-knowledge* property (in the sense that the argument provided here gives an explicit way of extracting the secret from P). Finally, we also want to informally argue for the *zero-knowledge* property: the reason why a verifier does not learn anything about x by running the above protocol is because V could have "simulated" the protocol execution in its own head, without interacting with the real prover. Now, if what V learns from this "simulation" (which does not use x) is exactly the same as what V learns from interacting with P, then the interaction with P cannot possibly leak any information about x. In particular, Schnorr protocol can be simulated in the following way: in the simulation one starts by choosing a random e and z, then computing $a = g^z h^{-e}$. It can be shown that the distribution of such a simulated transcript is identical to the distribution of (a, e, z) in a real execution of the protocol.[3]

Schnorr protocol, or variants of it, are widely used in practice, including as a building block in popular digital signature scheme such as (EC) DSA: such signature schemes are obtained by compiling a (variants of) Schnorr protocol using the Fiat-Shamir heuristic [FS86], which is a technique to make *public coin ZK protocols* (i.e., protocols where the verifier only samples a random challenge like

---

[3] More on this can be found in the many textbooks of lecture notes available on the topic e.g., [Dam02].

in Schnorr protocols) *non-interactive* in the *random oracle model*[4]: in a nutshell, in the Fiat-Shamir heuristic the challenge e is not chosen by the verifier but it is generated directly by the prover using a hash function on input a: this forces the prover to "commit" to a before receiving the challenge e, and therefore the prover cannot produce fake proofs (as a simulator could). Under the assumption that the hash function behaves like a random function, the challenge e is now chosen uniformly at random, exactly as a real verifier would, and therefore the security properties of the protocol are preserved.

Non-interactive ZK proofs constructed combining Schnorr protocol and the Fiat-Shamir heuristic can be easily turned into digital signatures schemes in the following way: x is the signing key and h is the public key. To sign a message m the signer constructs a (non-interactive) ZK proof where the challenge e is derived by hashing, in addition to a, the message m. Intuitively since only someone who knows the secret x can construct such a proof, and since the proof is linked to the message m, the verifier can be sure that P has seen and signed the message m.

1.3   Known Zero-Knowledge Protocols

Seminal results from the 80s tell us that *everything that is provable is provable in zero-knowledge*. In particular, not only any NP statement can be proven in ZK [GMW86], but even statements in IP can be proven in ZK as well [BGG+88]. Unfortunately these feasibility results use expensive Karp reductions and are therefore not particularly useful when trying to construct ZK protocols that are efficient enough to be used in practice.

The Schnorr protocol presented above is extremely efficient, and protocols with similar efficiency exist for all languages with enough "algebraic" structure. Following Schnorr work, a large body of literature has investigated the efficiency of ZK-protocols for proving relations between discrete logarithms, also over bilinear groups (e.g., the celebrated Groth-Sahai proofs [GS08]).

Unfortunately, when it comes to generic, non-algebraic statements such as "*I know x such that $y = h(x)$*" (for some concrete hash function h such as the SHA family, which is best expressed by a Boolean circuit) very few efficient protocols are known. A notable class of protocols which allow to prove generic statements are SNARKs: a SNARK (or *succint non-intearctive argument of knowledge*) allows to construct proofs which are very *short* and extremely *efficient to verify*. This has been proven true in practice by recent implementations of SNARKs such as libsnark [BCG+13, BCTV14] or Pinocchio [PHGR13, PHGR16]. SNARKs are perfect in situations where a proof needs to be verified by a large number of verifiers, such as in the cryptocurrency Zerocash [BCG+14]. Unfortunately the computational overhead for generating the proofs is quite high (due to the use of expensive public key operations for each gate in the circuit describing the function to be verified): for instance, for a concrete hash function such as SHA256,

---

[4] A good introduction to this somehow controversial model often used in cryptographic proofs can be found in [KL14].

the size of a SNARK is only a few hundred bytes and the verification time is in the order of few milliseconds. However, the proving time is in the order of seconds.

1.4 ZKGC: Zero-Knowledge from Garbled Circuits

The first protocol which allows to efficiently prove non-algebraic statements was proposed by Jawurek et al. [JKO13], and it is known as *ZKGC* or *zero-knowledge from garbled circuits*.

In a nutshell, garbled circuits are a cryptographic primitive which allows to evaluate *encrypted functions* on *encrypted inputs* while preserving useful security properties such as *privacy, authenticity, obliviusness* etc. [BHR12].

Garbled circuits were first introduced by Yao [Yao86] as a tool for implementing *secure two-party computation* or 2PC. In 2PC we have two parties, say A and B, who wish to compute a (publicly known) joint function f of their secret inputs x and y respectively. Intuitively, 2PC ensures that the only thing the two parties learn is the desired output $f(x, y)$ and nothing else about the secret input of the other party. Since the first public implementation of 2PC based on garbled circuit (the well known Fairplay system [MNPS04]), there has been a huge improvements in the performances of garbled circuits and 2PC protocols.

The starting point of ZKGC is a quite simply observation, namely that ZK is a proper subset of 2PC. In particular, ZK is the special case of 2PC in which only one of the parties has a secret input. Therefore it is natural to ask whether it is possible to optimize the existing (already very efficient) 2PC protocols to this specific setting.

The work of Jawurek et al. [JKO13] shows that this is indeed the case: first of all, the standard Yao's protocol for 2PC is *off-the-shelf* a *honest-verifies* ZK protocol i.e., a ZK protocol where the ZK property only holds against verifiers that follow the protocol correctly. The main problem with malicious verifiers (i.e., verifiers that might deviate from the protocol specification) is that a malicious verifier can garble an adversarially chosen function g instead of the function f agreed upon by the parties. This kind of malicious behaviour is undetectable: intuitively, since the protocol uses garbled circuits, a garbling of f and a garbling of g are indistinguishable in the eyes of the honest prover. Moreover, this can be easily used to break the ZK property: for example, $g(x)$ could be the function that leaks the most significant bit of x.

Yao's protocol for 2PC suffers from the same vulnerability: also here a malicious party can garble the wrong function and break the security of the protocol. There are several ways to deal with this in 2PC, but even the most efficient solution (e.g., the *cut-and-choose* approach in its most efficient instantiation [Lin13]) still requires to garble s copies of the function to get security 2^{-s}, meaning that in practice this incurs in computation and communication overhead of $s \geq 40$.

The approach taken in ZKGC is different: The main idea behind ZKGC is that in the special case of ZK the verifier has no input and therefore the verifier could reveal the randomness used to garble the function after the protocol execution without impacting security. This could in turn be used by the prover to check that the garbled function is indeed the one that was agreed upon. Of course this is not enough to achieve ZK, since it only allows to detect that a party has cheated after the information might have already been leaked.

This is fixed in ZKGC by letting the prover first commit to the output (i.e., the verifier does not learn anything yet thanks to the hiding property of the commitment scheme), then the verifier reveals the randomness used in the garbling (so that the prover can abort if a cheating attempt is detected), and finally the prover can open the commitment (and thanks to the binding property of the commitment the verifier is ensured that the output is the same as the one that the prover computed *before* the prover received the randomness of the garbling).

In conclusion ZKGC allows to construct ZK protocols with efficiency comparable to the passive secure version of Yao's protocol (while achieving security even against malicious provers and verifiers). In particular, this means that only a fixed number of public-key operations are needed (to run the *oblivious transfers* necessary during the input phase), and the protocol otherwise only uses a constant number of (cheaper) symmetric key operations per Boolean gate in the circuit of f. The details of the protocol can be found in [JKO13].

1.5   Privacy-Free Garbling Schemes

The ZKGC protocol can be made even more efficient using the following observation: in the specific ZK application one of the parties (the prover) knows the entire input, and therefore the prover also knows all the intermediate values for each wire in the Boolean circuit implementing f. This is in contrast with the 2PC setting in which each party only knows some of the input wires and therefore the intermediate values must be kept secret. It is therefore natural to ask whether one can construct more efficient garbling schemes which do not satisfy the privacy requirements (but still satisfy the *authenticity* requirement needed for ZK).

Frederiksen et al. [FNO15] answered this question in the affirmative by showing garbling schemes in which the evaluation algorithm is not "oblivious" but depends instead on the inputs to each gate. This allows significant savings in both the communication and the computation overhead of the garbling scheme. We refer to [FNO15] for more details on the constructions and their performances. Currently, the most efficient privacy-free garbling scheme is the one proposed by [ZRE15] which requires to transfer a single ciphertext for each AND gate in the circuit (and where linear gates e.g., XOR are "for free").

1.6   ZKBoo: Zero-Knowledge from Multiparty Computation

Ishai et al. [IKOS07,IKOS09] showed how to construct ZK protocols from *secure multiparty computation (MPC)* protocols. On top of creating a bridge between two fascinating topics in modern cryptography, this paper showed a number of asymptotically efficient ZK protocols which are obtained by instantiating their approach with the right (asymptotically) efficient MPC protocols. The question of whether this approach would lead to efficient ZK protocols in practice was left open.

The work of Giacomelli et al. [GMO16], known as ZKBoo, can be seen as a generalization, simplification and implementation of the proposal of Ishai et al. with focus on practical efficiency.

In a nutshell, to construct a ZKBoo proofs for a function f one first has to find a suitable *(2,3)-decomposition* of the function f: in a nutshell, this is a way of computing $f(x)$ by first splitting the input x into three shares $w_{1,1}, w_{1,2}, w_{1,3}$ such that $w_{1,1} \oplus w_{1,2} \oplus w_{1,3} = x$. Then, the computation of f proceeds in layers such that at each layer there are three functions $f_{i,1}, f_{i,2}, f_{i,3}$ such that $f_{i,j}$ takes input *only* $w_{i,j}$ and $w_{i,j+1}$ and produces some output $w_{i+1,j}$.[5] We call a decomposition *correct* if the output $y = f(x)$ can be reconstructed by XOR'ing the outputs of the last layer, and we call a decomposition private if for all $j \in \{1, 2, 3\}$, the values $\{(w_{i,j}, w_{i,j+1})\}_i$ can be simulated without knowledge of x. Such decompositions exist for any (Boolean or arithmetic) circuit (this technique is described in [GMO16] under the name *linear decomposition*).

Given such a decomposition we can construct a ZK protocol in the following way (note that the protocol has the same structure as the Schnorr protocol introduced before i.e., it is a Σ-protocol):

1. P computes $f(x)$ using the decomposition, then generates three (hiding and binding) commitments c_1, c_2, c_3 to the values $\{w_{i,j}\}_i$, and sends those commitments to V;
2. V chooses a random challenge $e \in \{1, 2, 3\}$;
3. P opens the commitments c_e and c_{e+1} revealing the values $\{(w_{i,e}, w_{i,e+1})\}_i$ to V;
4. V outputs accept iff the computation of all the values $w_{i,e}$ was performed correctly: note that the verifier can check this since $w_{i,e} = f_{i-1,e}(w_{i-1,e}, w_{i-1,e+1})$ i.e., all the computations in the decomposition only depend on two of the three values.

It can be shown that the protocol is sound (and can be made a proof of knowledge) due to the correctness of the decomposition and the binding property of the commitment (in particular the protocol has soundness error 2/3 and must therefore be repeated multiple times to achieve a negligible soundness error), and it can be shown that the protocol is zero-knowledge since the decomposition is private and the commitments are hiding. When compared with ZKGC, ZKBoo has two main advantages:

1. it does not use any public-key operations (it only uses commitment schemes which can be efficiently instantiated in practice using hash functions); and,
2. it is a *public-coin* protocol and therefore it can be made *non-interactive* using the Fiat-Shamir heuristic.

Using ZKBoo it is possible to construct very *fast* and *non-interactive* proofs for interesting Boolean circuits (such as hash functions in the SHA family).

---

[5] Modular reductions are implicit in the indices i.e., $3 + 1 = 1$.

In particular, the time to generate and verify a proof is in the order of milliseconds. On the negative side, the proofs generated by ZKBoo are quite large, in the order of hundreds of thousands of kilobytes for the SHA family. An improvement to ZKBoo, named ZKB++ was recently proposed [GCZ16]. This improved protocol produces proofs with size about a half of those produced by ZKBoo.

1.7   Digital Signatures from Symmetric Primitives

Two independent works by Derler et al. [DOR+16] and Goldfeder et al. [GCZ16] (later merged into Chase et al. [CDG+17]), proposed to construct digital signatures using ZKBoo/ZKB++ together with the Fiat-Shamir heuristic (using a similar approach to the one described earlier for the Schnorr protocol). In a nutshell, a signature scheme can now be constructed given any one-way function f: the secret key for the signature scheme is defined to be an input x, while the verification key is the image of x via the one-way function i.e., $y = f(x)$. To generate a signature the signer constructs a non-interactive ZKB++ proof of knowledge of the preimage x, where the challenge for the proof is derived using the Fiat-Shamir heuristic (and including the message to be signed).

To construct a signature scheme which is as efficient as possible using this approach one has to find a one-way function f which can be described using a Boolean circuit with a minimal number of AND gates. Fortunately, the design of such primitives has already been studied in the context of symmetric crypto primitives to be used in connection with MPC and homomorphic encryption, thus the choice fell on the LowMC cipher family [ARS+15, ARS+16].

An interesting property of the signature schemes obtained with this approach is that their security relies only on symmetric crypto primitives (block ciphers and hash functions). Therefore these signature schemes are a viable candidate for *post-quantum* signatures i.e., they can assumed to be secure also in the presence of quantum computers (as opposed to factoring or discrete log based signatures). See [CDG+17] for an extensive discussion on how these signatures compare with other post-quantum signature schemes.

1.8   Conclusions

As ZK protocols are one of the fundamental tools in modern cryptography, the availability of practically efficient ZK protocols is expected to enable a large number of applications. Several examples of this have already appeared in the literature, including: attribute based key exchange [KKL+16], enforcing input validity in 2PC [Bau16, KMW16, AMR17], ZK for RAM programs [HMR15, MRS17], anonymous credentials [CGM16], blind certificate authority registration [WPSR16], and more are expected to appear. The major open problem for this area of research is to significantly reduce the size of the proofs (which is currently the main bottleneck) without relying on computationally more expensive cryptographic primitives.

Acknowledgements. Research supported by the Danish Council for Independent Research, COST Action IC1306 and the European Union Horizon 2020 research and innovation programme under grant agreement No. 731583 (SODA).

References

[AMR17] Afshar, A., Mohassel, P., Rosulek, M.: Efficient maliciously secure two party computation for mixed programs. IACR Cryptology ePrint Archive, 2017:62 (2017)

[ARS+15] Albrecht, M.R., Rechberger, C., Schneider, T., Tiessen, T., Zohner, M.: Ciphers for MPC and FHE. In: Oswald, E., Fischlin, M. (eds.) EUROCRYPT 2015. LNCS, vol. 9056, pp. 430–454. Springer, Heidelberg (2015). doi:10.1007/978-3-662-46800-5_17

[ARS+16] Albrecht, M.R., Rechberger, C., Schneider, T., Tiessen, T., Zohner, M.: Ciphers for MPC and FHE. IACR Cryptology ePrint Archive, 2016:687 (2016)

[Bau16] Baum, C.: On garbling schemes with and without privacy. In: Zikas, V., De Prisco, R. (eds.) SCN 2016. LNCS, vol. 9841, pp. 468–485. Springer, Cham (2016). doi:10.1007/978-3-319-44618-9_25

[BCG+13] Ben-Sasson, E., Chiesa, A., Genkin, D., Tromer, E., Virza, M.: SNARKs for C: verifying program executions succinctly and in zero knowledge. In: Canetti, R., Garay, J.A. (eds.) CRYPTO 2013. LNCS, vol. 8043, pp. 90–108. Springer, Heidelberg (2013). doi:10.1007/978-3-642-40084-1_6

[BCG+14] Ben-Sasson E., Chiesa, A., Garman, C., Green, M., Miers, I., Tromer, E., Virza, M.: Zerocash: decentralized anonymous payments from bitcoin. In: 2014 IEEE Symposium on Security and Privacy (SP 2014), Berkeley, 18–21 May 2014, pp. 459–474 (2014)

[BCTV14] Ben-Sasson, E., Chiesa, A., Tromer, E., Virza, M.: Succinct non-interactive zero knowledge for a von Neumann architecture. In: Proceedings of the 23rd USENIX Security Symposium, San Diego, 20–22 August 2014, pp. 781–796 (2014)

[BGG+88] Ben-Or, M., Goldreich, O., Goldwasser, S., Håstad, J., Kilian, J., Micali, S., Rogaway, P.: Everything provable is provable in zero-knowledge. In: Goldwasser, S. (ed.) CRYPTO 1988. LNCS, vol. 403, pp. 37–56. Springer, New York (1990). doi:10.1007/0-387-34799-2_4

[BHR12] Bellare, M., Hoang, V.T., Rogaway, P.: Foundations of garbled circuits. In: The ACM Conference on Computer and Communications Security (CCS 2012), Raleigh, 16–18 October 2012, pp. 784–796 (2012)

[CDG+17] Chase, M., Derler, D., Goldfeder, S., Orlandi, C., Ramacher, S., Rechberger, C., Slamanig, D., Zaverucha, G.: Post-quantum zero-knowledge and signatures from symmetric-key primitives. In: CCS 2017. ACM (2017, to appear). http://eprint.iacr.org/2017/279

[CGM16] Chase, M., Ganesh, C., Mohassel, P.: Efficient zero-knowledge proof of algebraic and non-algebraic statements with applications to privacy preserving credentials. In: Robshaw, M., Katz, J. (eds.) CRYPTO 2016. LNCS, vol. 9816, pp. 499–530. Springer, Heidelberg (2016). doi:10.1007/978-3-662-53015-3_18

[Dam02] Damgård, I.: On σ-protocols. Lecture Notes, University of Aarhus, Department for Computer Science (2002)

[DOR+16] Derler, D., Orlandi, C., Ramacher, S., Rechberger, C., Slamanig, D.: Digital signatures from symmetric-key primitives. Cryptology ePrint Archive, Report 2016/1085 (2016). http://eprint.iacr.org/2016/1085

[FNO15] Frederiksen, T.K., Nielsen, J.B., Orlandi, C.: Privacy-free garbled circuits with applications to efficient zero-knowledge. In: Oswald, E., Fischlin, M. (eds.) EUROCRYPT 2015. LNCS, vol. 9057, pp. 191–219. Springer, Heidelberg (2015). doi:10.1007/978-3-662-46803-6_7

[FS86] Fiat, A., Shamir, A.: How to prove yourself: practical solutions to identification and signature problems. In: Odlyzko, A.M. (ed.) CRYPTO 1986. LNCS, vol. 263, pp. 186–194. Springer, Heidelberg (1987). doi:10.1007/3-540-47721-7_12

[GCZ16] Goldfeder, S., Chase, M., Zaverucha, G.: Efficient post-quantum zero-knowledge and signatures. Cryptology ePrint Archive, Report 2016/1110 (2016). http://eprint.iacr.org/2016/1110

[GMO16] Giacomelli, I., Madsen, J., Orlandi, C.: ZKBoo: faster zero-knowledge for Boolean circuits. In: 25th USENIX Security Symposium (USENIX Security 2016), Austin, 10–12 August 2016, pp. 1069–1083 (2016)

[GMR85] Goldwasser, S., Micali, S., Rackoff, C.: The knowledge complexity of interactive proof-systems (extended abstract). In: Proceedings of the 17th Annual ACM Symposium on Theory of Computing, 6–8 May 1985, Providence, pp. 291–304 (1985)

[GMR89] Goldwasser, S., Micali, S., Rackoff, C.: The knowledge complexity of interactive proof systems. SIAM J. Comput. 18(1), 186–208 (1989)

[GMW86] Goldreich, O., Micali, S., Wigderson, A.: How to prove all NP statements in zero-knowledge and a methodology of cryptographic protocol design (extended abstract). In: Odlyzko, A.M. (ed.) CRYPTO 1986. LNCS, vol. 263, pp. 171–185. Springer, Heidelberg (1987). doi:10.1007/3-540-47721-7_11

[Gol01] Goldreich, O.: The Foundations of Cryptography. Basic Techniques, vol. 1. Cambridge University Press, Cambridge (2001)

[Gol04] Goldreich, O.: The Foundations of Cryptography. Basic Applications, vol. 2. Cambridge University Press, Cambridge (2004)

[GS08] Groth, J., Sahai, A.: Efficient non-interactive proof systems for bilinear groups. In: Smart, N. (ed.) EUROCRYPT 2008. LNCS, vol. 4965, pp. 415–432. Springer, Heidelberg (2008). doi:10.1007/978-3-540-78967-3_24

[HMR15] Hu, Z., Mohassel, P., Rosulek, M.: Efficient zero-knowledge proofs of non-algebraic statements with sublinear amortized cost. In: Gennaro, R., Robshaw, M. (eds.) CRYPTO 2015. LNCS, vol. 9216, pp. 150–169. Springer, Heidelberg (2015). doi:10.1007/978-3-662-48000-7_8

[IKOS07] Ishai, Y., Kushilevitz, E., Ostrovsky, R., Sahai, A.: Zero-knowledge from secure multiparty computation. In: Proceedings of the 39th Annual ACM Symposium on Theory of Computing, San Diego, 11–13 June 2007, pp. 21–30 (2007)

[IKOS09] Ishai, Y., Kushilevitz, E., Ostrovsky, R., Sahai, A.: Zero-knowledge proofs from secure multiparty computation. SIAM J. Comput. 39(3), 1121–1152 (2009)

[JKO13] Jawurek, M., Kerschbaum, F., Orlandi, C.: Zero-knowledge using garbled circuits: how to prove non-algebraic statements efficiently. In: 2013 ACM SIGSAC Conference on Computer and Communications Security (CCS 2013), Berlin, 4–8 November 2013, pp. 955–966 (2013)

[KKL+16] Kolesnikov, V., Krawczyk, H., Lindell, Y., Malozemoff, A.J., Rabin, T.: Attribute-based key exchange with general policies. In: Proceedings of the 2016 ACM SIGSAC Conference on Computer and Communications Security, Vienna, 24–28 October 2016, pp. 1451–1463 (2016)

[KL14] Katz, J., Lindell, Y.: Introduction to Modern Cryptography, 2nd edn. CRC Press, Boca Raton (2014)

[KMW16] Katz, J., Malozemoff, A.J., Wang, X.S.: Efficiently enforcing input validity in secure two-party computation. IACR Cryptology ePrint Archive, 2016:184 (2016)

[Lin13] Lindell, Y.: Fast cut-and-choose based protocols for malicious and covert adversaries. In: Canetti, R., Garay, J.A. (eds.) CRYPTO 2013. LNCS, vol. 8043, pp. 1–17. Springer, Heidelberg (2013). doi:10.1007/978-3-642-40084-1_1

[MNPS04] Malkhi, D., Nisan, N., Pinkas, B., Sella, Y.: Fairplay - secure two-party computation system. In: Proceedings of the 13th USENIX Security Symposium, San Diego, 9–13 August 2004, pp. 287–302 (2004)

[MRS17] Mohassel, P., Rosulek, M., Scafuro, A.: Sublinear zero-knowledge arguments for RAM programs. In: Coron, J.-S., Nielsen, J.B. (eds.) EUROCRYPT 2017. LNCS, vol. 10210, pp. 501–531. Springer, Cham (2017). doi:10.1007/978-3-319-56620-7_18

[PHGR13] Parno, B., Howell, J., Gentry, C., Raykova, M.: Pinocchio: nearly practical verifiable computation. In: 2013 IEEE Symposium on Security and Privacy (SP 2013), Berkeley, 19–22 May 2013, pp. 238–252 (2013)

[PHGR16] Parno, B., Howell, J., Gentry, C., Raykova, M.: Pinocchio: nearly practical verifiable computation. Commun. ACM **59**(2), 103–112 (2016)

[Sch89] Schnorr, C.P.: Efficient identification and signatures for smart cards. In: Brassard, G. (ed.) CRYPTO 1989. LNCS, vol. 435, pp. 239–252. Springer, New York (1990). doi:10.1007/0-387-34805-0_22

[WPSR16] Wang, L., Pass, R., Shelat, A., Ristenpart, T.: Secure channel injection and anonymous proofs of account ownership. IACR Cryptology ePrint Archive, 2016:925 (2016)

[Yao86] Yao, A.C.-C.: How to generate and exchange secrets (extended abstract). In: 27th Annual Symposium on Foundations of Computer Science, Toronto, 27–29 October 1986, pp. 162–167 (1986)

[ZRE15] Zahur, S., Rosulek, M., Evans, D.: Two halves make a whole. In: Oswald, E., Fischlin, M. (eds.) EUROCRYPT 2015. LNCS, vol. 9057, pp. 220–250. Springer, Heidelberg (2015). doi:10.1007/978-3-662-46803-6_8

Stochastic Side-Channel Leakage Analysis *via* Orthonormal Decomposition

Sylvain Guilley[1,2]([✉]), Annelie Heuser[3], Tang Ming[4], and Olivier Rioul[2]

[1] Secure-IC S.A.S., Cesson-Sévigné, France
`sylvain.guilley@secure-ic.com`
[2] Telecom-ParisTech, LTCI, Université Paris-Saclay, Paris, France
[3] CNRS, IRISA, Rennes, France
[4] Wuhan University, Wuhan, China

Abstract. Side-channel attacks of maximal efficiency require an accurate knowledge of the leakage function. Template attacks have been introduced by Chari et al. at CHES 2002 to estimate the leakage function using available training data. Schindler et al. noticed at CHES 2005 that the complexity of profiling could be alleviated if the evaluator has some prior knowledge on the leakage function. The initial idea of Schindler is that an engineer can model the leakage from the structure of the circuit. However, for some thin CMOS technologies or some advanced countermeasures, the engineer intuition might not be sufficient. Therefore, inferring the leakage function based on profiling is still important. In the state-of-the-art, though, the profiling stage is conducted based on a linear regression in a non-orthonormal basis. This does not allow for an easy interpretation because the components are not independent.

In this paper, we present a method to characterize the leakage based on a Walsh-Hadamard orthonormal basis with staggered degrees, which allows for direct interpretations in terms of bits interactions. A straightforward application is the characterization of a class of devices in order to understand their leakage structure. Such information is precious for designers and also for evaluators, who can devise attack bases relevantly.

Keywords: Side-channel analysis · Stochastic attacks · Leakage model · Pseudo-Boolean functions · Orthonormal bases · Leakage characterization

1 Introduction

The existence of side-channels weakens the security of embedded devices, as it allows an attacker to retrieve information about secret keys. The best attacks require the best possible knowledge about the leakage function. A first method in this direction consists of exhaustive characterizations, referred to as *templates* by Chari et al. [5]. Templates are asymptotically perfect estimations of the model, but as pointed out by Schindler [15], they may be inaccurate when there is only a limited amount of profiling traces. Therefore, Schindler has suggested to simplify

© Springer International Publishing AG 2017
P. Farshim and E. Simion (Eds.): SecITC 2017, LNCS 10543, pp. 12–27, 2017.
https://doi.org/10.1007/978-3-319-69284-5_2

the characterization using *stochastic* attacks. While the template method consists in profiling leakage values for all configurations of intermediate variables, which Schindler describes as a projection over a full basis, stochastic attacks consist in characterizing the leakage over a basis of smaller dimensionality.

Leakage characterization does not only benefit to actual attacks. As shown by Kasper et al. [11], it is also a *constructive* feature: when the basis is able to describe the switching activity of the circuit, the estimated weights (basis coefficients) highlight specific exploitable security flaws in the implementation. In their case study, the absolute value of the weight corresponding to one specific bit showed that is was leaking in an outstanding way, and this could be connected to the underlying hardware components (that bit was driving a multiplexer network).

Another motivation is for implementing masking countermeasures. The sensitive data is split into shares which should not interfere physically. Stochastic characterization of the leakage of a *bit pairs* (and in general, of a *bit tuples*) belonging to different shares can reveal flaws in the implementation.

Additionally, stochastic characterization can also benefit to the analysis of unprotected implementations. Recent works showed that, if the *linear basis* describing the switching activity of each bit independently is extended to a *nonlinear basis* which also includes *interactions* between bits, then attacks are more successful in terms of success rate (see e.g., [8,13]). Interestingly, while we know that the consideration of nonlinear bases improves the attack, no sound explanations have been given about what precise information is captured by these nonlinear basis vectors. In [10,13] the authors mention *cross-talk* and *glitch* effects as one possible reason. Up to now, these effects could not be precisely accounted for. One possible reason is that a badly chosen nonlinear basis extension, made with products of bits (i.e., *monomials*), is neither normalized nor orthogonal. As a result, the estimated weights cannot be compared to each other and it seems difficult to draw conclusions about the influence of either individual bits or bit interactions. While the basis *normalization* can be easily carried out (see e.g., [10]), any unstructured *orthogonalization* procedure comes at the expense of the loss of its interpretability in terms of bit interactions, due to the underlying complex change of basis.

Contributions. The goal of this paper is to describe the best possible basis decomposition that is able to isolate leakage from a given coupling of pairs, triples, . . . , tuples of bits, independently of the others. We conduct an extensive study of the underlying basis and find a surprisingly simple method to compute the orthonormalized basis. Our method does not only give a feasible solution to interpret the results but it also helps avoid stability problems that occur using standard procedures [16, Sect. 4.2]. The practicability of our methods is tested using simulations and measurements where a leakage is attributed to a tuple of interacting bits.

Outline. The remainder of the paper is organized as follows. Section 2 provides mathematical background for stochastic profiling. Our contribution starts at

Sect. 3, where we derive a novel basis for leakage function decomposition which allows for an easy interpretation in terms of degrees. The method consists in applying a Gram-Schmidt transform on the monomial basis, ordered according to monomial degrees. In Sect. 4 we investigate the leakage estimation in the new basis, together with a fast computation based on the Fourier transform. Practical validation on simulated and real-world traces is shown in Sect. 5. Finally Sect. 6 concludes. Appendix A shows how to estimate projections, and gives an exemple of a "bad" projection into a non-orthogonal basis.

2   Stochastic Profiling

2.1   Leakage Model

Consider a leaking device which manipulates some secret key k. The cryptographic operations involve xoring k with some (plain or cipher) text T. The attacker focuses on manageable parts of the text and key, and T is taken as an n-bit byte (typically $n = 8$). Thus the leakage function f applies to $T \oplus k$ together with some additive noise N, modeled as a normal random variable $N \sim \mathcal{N}(0, \sigma^2)$. The resulting leakage X is given by the equation

$$X = f(T \oplus k) + N. \tag{1}$$

The purpose of this paper is to characterize f which maps the finite set $\mathbb{F}_2^n = \{0, 1\}^n$ to the set of real numbers $\mathbb{R}$. A simple example would be the Hamming weight $f = w_H$. Often, f is taken as the composition of some cryptographic function, such as a substitution box $S : \{0, 1\}^n \to \{0, 1\}^n$, and a leakage function, such as the Hamming weight w_H. This is represented in Fig. 1. In practice, the mapping from $S(T \oplus k) \in \{0, 1\}^n$ to $\mathbb{R}$ can be more complex.

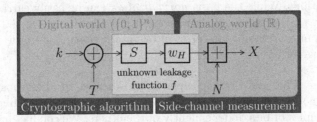

Fig. 1. Setup considered in this paper: f is the unknown

In the following, we consider several independent and identically distributed (i.i.d.) realizations of T, N and X, denoted by $(t_1, \ldots, t_Q) = (t_q)_{q \in \{1, \ldots, Q\}}$, $(n_q)_{q \in \{1, \ldots, Q\}}$ and $(x_q)_{q \in \{1, \ldots, Q\}}$, respectively, where Q denotes the number of queries.

2.2   Notations for Sums and Products

Sum notations will differ depending on whether the considered variables lie in $\mathbb{F}_2^n$ or $\mathbb{R}$. Let $t \in \mathbb{F}_2^n$ be any n-bit vector with bits $t_0, t_1, \ldots, t_{n-1}$. We let $t_i \oplus t_j$ be the exclusive-or addition of bits t_i and t_j in $\mathbb{F}_2$, such that $1 \oplus 1 = 0$, while the usual sum notation $t_i + t_j$ refers to the addition in $\mathbb{R}$, where $1 + 1 = 2$. For the product, there is no such complication. Letting $\wedge$ be the 'and' operator for multiplication in $\mathbb{F}_2$ and $\times$ be the usual multiplicative product in $\mathbb{R}$, we have in fact $t_i \wedge t_j = t_i \times t_j$ for any two bits t_i and t_j in $\{0, 1\}$. Therefore, we will simply denote this product by $t_i t_j$, and use the notation $\prod_{i=0}^{n-1} t_i$ to denote the conjunction of all bits of bit vector t.

2.3   Template and Stochastic Attacks

Template attacks [5] consist in an offline estimation of Eq. (1) for all values t of realizations of T and all choices of the secret key k. This profiling phase is followed by an online application of the maximum likelihood principle to uncover the unknown key. However, template attacks cannot provide an analytic characterization of the leakage. For instance, templates cannot answer the question: "*are bits 2 and 3 of T leaking together?*". We will show in Fig. 4(b) and (c) that our leakage characterization can give a quantitative answer.

While template attacks are *data-driven*, stochastic attacks are *model-driven*: They assume authoritatively that Eq. (1) can be considered to belong to a specific subset of functions $\mathbb{F}_2^n \to \mathbb{R}$. However, the classical approach is to assume some basis for f based on the engineer's intuition. In contract, we aim to find a method to select the most suitable basis for the representation of f.

2.4   Bases and Orthonormality

Let $\mathcal{E}$ be the set of so-called *pseudo-Boolean* [4, Sect. 2.1] functions $\mathbb{F}_2^n \to \mathbb{R}$, which forms a Euclidean vector space over $\mathbb{R}$ of dimension 2^n. The *scalar product* between two vectors f_0 and f_1 in $\mathcal{E}$ is $\langle f_0 | f_1 \rangle = \sum_{t \in \mathbb{F}_2^n} f_0(t) f_1(t)$ and the corresponding *norm* is $\|f\|_2 = \sqrt{\langle f | f \rangle}$. Any linearly independent family of 2^n vectors $(\psi_u)_{u \in \mathbb{F}_2^n}$ form a *basis* of $\mathcal{E}$. This basis is *orthonormal* if $\langle \psi_u | \psi_v \rangle = 0$ for all $u \neq v$ and $= 1$ if $u = v$. In this case an arbitrary pseudo-Boolean function $f \in \mathcal{E}$ can be written as the sum of orthogonal projections

$$f = \sum_{u \in \mathbb{F}_2^n} a_u \psi_u \quad \text{where} \quad a_u = \langle f | \psi_u \rangle \in \mathbb{R} \tag{2}$$

The leakage function $f : \mathbb{F}_2^n \to \mathbb{R}$ is an element of $\mathcal{E}$ that we would like to characterize through a convenient vector basis of $\mathcal{E}$. Two requirements are:

- the basis should somehow relate to bit combinations to make an easy interpretation of the leakage structure in terms of the interactions between bits;
- the basis should be orthonormal so that the characterization of each basis vector is uncorrelated to the other basis vectors.

Appendix A provides an analysis which explains why the use of a non-orthogonal basis is misleading for the interpretation of bit interactions. Appendix A.1 details how coordinates in an orthonormal basis can be estimated with a *correlation method*, and Appendix A.2 shows that the blind application of this method to a non-orthogonal basis yields erroneous results.

2.5   Canonical and Monomial Bases; Degree

The *canonical basis* $(\delta_u)_{u \in \mathbb{F}_2^n}$ of $\mathcal{E}$ is defined by

$$\delta_u(t) = \prod_{i=0}^{n-1} (t_i \oplus u_i) = \begin{cases} 1 & \text{if } t = u, \\ 0 & \text{otherwise,} \end{cases}$$

while the *monomial basis* $(\phi_u)_{u \in \mathbb{F}_2^n}$ of $\mathcal{E}$ is defined by

$$\phi_u(t) = \prod_{i | u_i = 1} t_i = \prod_{i=0}^{n-1} t_i^{u_i}. \tag{3}$$

where the power notation is simply $t_i^0 = 1$ and $t_i^1 = t_i$.

Definition 1 (Degree). *The* degree *of the monomial* $\phi_u(t) = \prod_{i=0}^{n-1} t_i^{u_i}$ *is the number of coordinates involved in the product, that is, the Hamming weight* $w_H(u) = \sum_{i=0}^{n-1} u_i$ *of* u.

The degree $\deg(f)$ *of any pseudo-Boolean function* $f : \mathbb{F}_2^n \to \mathbb{R}$ *is the maximum value of the degrees of the monomials* ϕ_u *in the decomposition of* f *over the monomial basis.*

A function of unit degree is simply a linear combination of bit values, also referred to as Unevenly Weighted Sum of Bits (UWSB) in the side-channel literature [9,17]. A function of degree >1 has *interacting bits* in its decomposition. For example, when the degree is two, product of bits $t_i t_j$ for $i \neq j$ are involved. The degree represents the maximum number of interacting bits.

2.6   Why Canonical and Monomial Bases Are Not Suitable

Properties of the canonical and monomial bases in terms of orthogonality and degree are as follows.

Lemma 2. *The canonical basis is orthonormal, but all vectors have degree* n.

Proof. Clearly $\|\delta_u\| = 1$ and $\langle \delta_u | \delta_v \rangle$ vanishes when $u \neq v$ since the supports of δ_u and δ_v are disjoint. This shows orthonormality. Regarding the degree, we have, for all $t, u \in \mathbb{F}_2^n$:

$$\delta_u(t) = \prod_{i/u_i=1} t_i \prod_{j/u_j=0} (1 - t_j).$$

Expending this sum we see that it includes the term $(+1)^{w_H(u)}(-1)^{n-w_H(u)}$ $\phi_{(1,\ldots,1)}$, where $(1, \ldots, 1)$ is the all-one n-bit vector. Since the latter has Hamming weight equal to n, the corresponding $\phi_{(1,\ldots,1)}$, and so δ_u, has degree n. $\qquad \square$

As a consequence, the canonical functions δ_u, albeit simple, are not of practical interest since being all of degree n they are not easily interpretable in terms of "interactions between bits".

On the other hand, the monomial basis is considered in the seminal paper on stochastic side-channel analysis by Schindler *et al.* [15, Eq. (23)], and is customary in side-channel analysis and well understood by engineers because the basis functions have staggered degrees $0, 1, \ldots, n$: While ϕ_0 is the constant 1, the basis vector ϕ_u simply represents the interactions between those bits t_i for which $u_1 = 1$. These basis functions, however, are not even orthogonal:

Lemma 3. *Any monomial basis function ϕ_u has degree equal to $w_H(u) \in \{0, 1, \ldots, n\}$, but the monomial basis is not orthonormal (not even orthogonal):*

$$\langle \phi_u | \phi_v \rangle = 2^{n - w_H(u \vee v)}$$

where $u \vee v$ denotes the bitwise inclusive 'or' of u and v.

Proof. By definition $\deg(\phi_u) = w_H(u)$. We have

$$\langle \phi_u | \phi_v \rangle = \sum_t \phi_u(t)\phi_v(t) = \sum_{t_0, \ldots, t_{n-1}} \prod_{i=0}^{n-1} t_i^{u_i + v_i} \tag{4}$$

$$= \prod_{i=0}^{n-1} \left(\sum_{t_i} t_i^{u_i + v_i} \right) = \prod_{i | u_i = v_i = 0} 2 \tag{5}$$

which is always nonzero. □

3   Orthonormalizing the Monomial Basis

The monomial basis is ordered by increasing degree (or Hamming weight). For exemple for $n = 3$, the basis vectors are enumerated in the following *weighting order*: $\phi_{(0,0,0)}$, $\phi_{(1,0,0)}$, $\phi_{(0,1,0)}$, $\phi_{(0,0,1)}$, $\phi_{(1,1,0)}$, $\phi_{(1,0,1)}$, $\phi_{(0,1,1)}$ and $\phi_{(1,1,1)}$. Vectors of same weight represent the same number of interacting bits. We proceed to carry out an orthonormalization process that preserves the weight ordering.

3.1   Gram-Schmidt Orthonormalization in Weighting Order

The new orthonormal basis ordered by degree is obtained from the monomial basis by the well-known *Gram-Schmidt orthonormalization*, yielding an orthonormal basis $(\psi_u)_{u \in \mathbb{F}_2^n}$ which can be constrained to *preserve the degree* (as we shall prove in Proposition 4). Algorithm 1 below is Gram-Schmidt procedure operating on vectors ϕ_u with u sorted by weighting order. We write interchangeably $u = (u_0, \ldots, u_{n-1}) \in \mathbb{F}_2^n$ and its equivalent $u = \sum_{i=0}^{n-1} u_i 2^i$ in $\{0, \ldots, 2^n - 1\}$. As the set $\{0, \ldots, 2^n - 1\}$ is totally ordered, this induces the natural *lexicographical* order on $\mathbb{F}_2^n$.

Input   : $(\phi_u)_{u \in \mathbb{F}_2^n}$, a basis of $\mathcal{E}$
Output : $(\psi_u)_{u \in \mathbb{F}_2^n}$, an orthonormal basis of $\mathcal{E}$

```
// Creation of the weighting order ....................................
1  W ← ∅
2  for w = 0 to n do
3  │   for j = 0 to 2^n - 1 do
4  │   │   if w_H(j) = w then
5  │   │   └   W ← W ∪ {j}
```

```
// Orthonormalization using Gram-Schmidt process ......................
6  for j = 0 to 2^n - 1 do
```
$$7 \quad \xi_{W[j]} \leftarrow \phi_{W[j]} - \sum_{i=0}^{j-1} \frac{\langle \phi_{W[j]} | \xi_{W[i]} \rangle}{\langle \xi_{W[i]} | \xi_{W[i]} \rangle} \xi_{W[i]}$$
$$8 \quad \psi_{W[i]} \leftarrow \frac{\xi_{W[j]}}{||\xi_{W[j]}||_2}$$
```
9  return (ψ_u)_{u ∈ F_2^n}
```

Algorithm 1. Gram-Schmidt orthonormalization in weighting order

Proposition 4 (Degree Preservation of the Gram-Schmidt Orthonormalization in Weighting Order). *Let $(\phi_u)_{u \in \mathbb{F}_2^n}$ be a basis of $\mathcal{E}$, such that $\deg(\phi_u) \leq \deg(\phi_v)$ if u is smaller than v with respect to the weighting order (that is $w_H(u) \leq w_H(v)$). Then the Gram-Schmidt orthonormalization process in weighting order (Algorithm 1) applied on $(\phi_u)_{u \in \mathbb{F}_2^n}$ yields a new basis $(\psi_u)_{u \in \mathbb{F}_2^n}$ where $\deg(\psi_u) = \deg(\phi_u)$, for all $u \in \mathbb{F}_2^n$.*

Proof. The weighting order is computed in Algorithm 1 between its lines 1 and 5. It consists in a permutation W of $\{0, \ldots, 2^n - 1\}$, which is such that:

$$\forall j, j' \in \{0, \ldots, 2^n - 1\}, \quad j \leq j' \implies w_H(W[j]) \leq w_H(W[j']). \tag{6}$$

In Algorithm 1, the first vector fetched from the monomial basis is ϕ_0, which has degree zero. Thus, the degree of $\psi_0 = \phi_0 / ||\phi_0||_2$ is also zero. Then, by induction on the loop index j (see line 6 of Algorithm 1), we see that the degree of $\psi_{W[j]}$ is equal to that of $\phi_{W[j]}$. Indeed:

- at line 7, we see that $\xi_{W[j]}$ is equal to $\phi_{W[j]}$ minus terms of lower (or equal) degree, owing to the weighting ordering of $W[j]$ (recall Eq. (6));
- at line 8, we see that the degree of $\psi_{W[j]}$ is the same as that of $f_{W[j]}$, because $\psi_{W[j]}$ is the unitary scaling of f_j, operation which keeps the degree unchanged.
□

The application of Algorithm 1 on $(\phi_u)_{u \in \mathbb{F}_2^n}$ thus yields a new basis $(\psi_u)_{u \in \mathbb{F}_2^n}$ which meets our requirements: it is orthonormal and ordered by degree.

3.2   Link to Walsh-Hadamard Matrix or Fourier Transform

The Walsh-Hadamard matrices of dimension 2^n for $n \in \mathbb{N}^+$ are given by the recursive formula:

$$H(2^n) = \begin{bmatrix} +H(2^{n-1}) & +H(2^{n-1}) \\ +H(2^{n-1}) & -H(2^{n-1}) \end{bmatrix} \qquad (n > 1)$$

where the lowest order of Walsh-Hadamard matrix is

$$H(2) = \begin{bmatrix} +1 & +1 \\ +1 & -1 \end{bmatrix}.$$

A matrix built according to this definition is also referred to as a *lexicographical ordered Walsh-Hadamard matrix*. Walsh-Hadamard matrices are specific square matrices with dimensions of some power of 2, entries of ± 1, and the property that the dot product of any two distinct rows (or columns) is zero.

It is well known that the Walsh-Hadamard matrix H_n is of the form $H_n = 2^{n/2}(\psi_u(t))_{u \in \mathbb{F}_2^n, t \in \mathbb{F}_2^n}$, where u and t are listed in *lexicographical order* (that is, $u \in \mathbb{F}_2^n$ ordered by increasing values of $\sum_{i=0}^{n-1} u_i 2^i$), and where

$$\psi_u(t) = \frac{1}{2^{n/2}}(-1)^{u \cdot t}$$

(where $u \cdot t = \bigoplus_{i=0}^{n-1} u_i t_i$ is the dot product of bitvectors u and t) forms a basis of $\mathcal{E}$ known as the *Fourier basis*.

Theorem 5 (Main Theoretical Result of the Paper). *The basis* $(\psi_u)_{u \in \mathbb{F}_2^n}$, *obtained by Algorithm 1 from the monomial basis* $(\phi_u)_{u \in \mathbb{F}_2^n}$, *coincides with the Fourier basis.*

Proof. Let $u \in \mathbb{F}_2^n$. We have that

$$\psi_u(t) = \frac{1}{2^{n/2}}(-1)^{u \cdot t} = \frac{1}{2^{n/2}} \prod_{i=0}^{n-1}(1 - 2t_i)^{u_i}.$$

The development of the product yields a sum of monomials of degrees at most $w_H(u)$. The (only) monomial of degree $w_H(u)$ is $c\phi_u(t)$, where the constant c is equal to $\frac{1}{2^{n/2}}(-2)^{w_H(u)}$. Thus, we have that:

$$\psi_u(t) = c\phi_u(t) - \underbrace{\text{monomials of degree strictly smaller than that of } \psi_u}.$$

orthogonal projection of ϕ_u on $\psi_{u'}$,
for each u' is smaller than u in the weighting order.

This is exactly the procedure of the Gram-Schmidt orthonormalization process in weighting order (line 7 in Algorithm 1). $\qquad\square$

Therefore, we have proven that using the Fourier basis $(\psi_u)_{u \in \mathbb{F}_2^n}$ for the projection of the leakage function, the evaluator keeps the mapping between:

– the basis vector $\psi_u : t \mapsto \frac{1}{2^{n/2}}(-1)^{u \cdot t}$, and
– the bit lines which interact (namely, the bits $\{0 \le i < n, \text{ s.t. } u_i = 1\}$).

Therefore, the leakage can be directly interpreted from the orthonormal projection of the leakage on ψ_u. and the corresponding coefficients a_u of $f : \mathbb{F}_2^n \to \mathbb{R}$ are those on the Fourier basis:

$$f(t) = \sum_u \langle f | \psi_u \rangle \psi_u(t) = \frac{1}{2^{n/2}} \sum_u a_u (-1)^{t \cdot u} \qquad \text{(Eq. (2) in Fourier basis),} \quad (7)$$

which is a *Fourier transform*. The coefficients a_u can be recovered as:

$$a_u = \frac{1}{2^{n/2}} \sum_t f(t)(-1)^{t \cdot u}, \qquad (8)$$

which is the corresponding *inverse Fourier transform*. Notice that direct (Eq. (7)) and inverse (Eq. (8)) Fourier transforms are the same in characteristic two (because $\forall u \in \mathbb{F}_2^n, -u = u$); put differently, the Fourier transform is involutive.

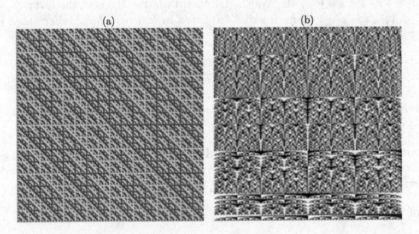

Fig. 2. (a) Walsh-Hadamard 256×256 matrix representation, (b) Truth table of Fourier basis (multiplied by $\sqrt{256} = 16$), in weighting order.

Application to the Case $n = 8$. In the case of byte-oriented block ciphers, such as the AES, the manipulated data are bytes of $n = 8$ bits. The $H(256)$ Walsh-Hadamard matrix is illustrated in Fig. 2(a). Dark pixels are -1 whereas white pixels are $+1$ values. The truth table of the Fourier basis (without the scaling factor of $2^{-n/2}$), represented in weighting order, is depicted in Fig. 2(b). This second matrix is simply the Walsh-Hadamard matrix where lines have been permuted to match the weighting order. One can see that the $H(256)$ matrix is symmetrical. In contrast, the truth table of the Fourier basis is structured as 9 horizontal stripes, comprising 1 (resp. 8, 28, 56, 70, 56, 28, 8 and 1) lines, corresponding to Hamming weight 0 (resp. 1, 2, 3, 4, 5, 6, 7 and 8). It is not immediate visually from Fig. 2(b) that the projection vectors have the same degrees in each "stripe".

3.3   Attribution of Leakage Using the Fourier Basis

Owing to the above properties, the attribution of the leakage using Fourier basis is straightforward:

- build a bitvector $u \in \{0,1\}^n$ where the bits $= 1$ are those we intend to test the interaction in terms of leakage. For instance, to extract the amount of leakage of the Least Significant Bit (LSB), use $u = (1,0,0,\ldots,0)$. Or to test the joint amount of leakage of bits 0 and 1, use $u = (1,1,0,\ldots,0)$;
- compute the projection of the leakage on vector ψ_u (see next section for an estimation method).

4   Estimation of the Projection onto the Fourier Basis

4.1   Exact Solution for the Estimation of the Basis Coefficients

Suppose we have Q leakage values $(x_1, \ldots, x_Q) \in \mathbb{R}^Q$ and let $a = (a_u)_{u \in \mathbb{F}_2^n} \in \mathbb{R}^{2^n}$ be the basis coefficients to be found. Due to the Gaussian nature of the noise, the minimum likelihood determination of a is the following convex optimization problem [10], which happens to be a linear regression problem:

$$\min_{a \in \mathbb{R}^{2^n}} \sum_{q=1}^{Q} \left(x_q - 2^{-n/2} \sum_{u \in \mathbb{F}_2^n} a_u (-1)^{u \cdot (t_q \oplus k)} \right)^2 = \min_{a \in \mathbb{R}^{2^n}} ||x - aG||^2, \qquad (9)$$

where in this case $|| \cdot ||$ is the norm-2 over $\mathbb{R}^Q$, and where G is a $2^n \times Q$ matrix, whose elements are $G[u,q] = 2^{-n/2}(-1)^{u \cdot (t_q \oplus k)}$.

Proposition 6. *The optimal value in Eq. (9) is* $a = xG^{\mathsf{T}}(GG^{\mathsf{T}})^{-1}$.

Proof. This is standard; see [1].

4.2   Fast (Approximate) Solution for the Estimation of $(a_u)_{u \in \mathbb{F}_2^n}$

The expression of Proposition 6 is well known to be a *Moore-Penrose pseudo-inverse*, see e.g. [16, p. 491]. However, it has never been explained in the field of side-channel analysis that the coefficients a_u can be estimated with the following fast formula (in the limit of the low of large numbers), which is an (inverse) *Fourier transform*:

Theorem 7 (Second Main Result of the Paper). *Given* Q *traces* $(x_1, \ldots, x_Q)$ *and the* Q *corresponding texts* $(t_1, \ldots, t_Q)$, *where the texts are assumed uniformly distributed over* $\mathbb{F}_2^n$, *the estimation of* a_u *in the law of large numbers is:*

$$a_u \approx \frac{2^{n/2}}{Q} \sum_{t \in \mathbb{F}_2^n} \left(\sum_{q/t_q = t} x_q \right) (-1)^{u \cdot (t \oplus k)} \qquad \text{when } Q \to \infty. \qquad (10)$$

Proof. Let us notice that xG^T is a vector of length 2^n, whose value at index $u \in \{0,1\}^n$ is $2^{-n/2} \sum_{q=1}^{Q} x_q(-1)^{u \cdot (t_q \oplus k)}$. Using the reordering of sums put forward in [12], this quantity is also $2^{-n/2} \sum_{t \in \mathbb{F}_2^n} \left(\sum_{q/t_q=t} x_q \right) (-1)^{u \cdot (t \oplus k)}$. Now, assuming that T is uniformly distributed on $\{0,1\}^n$, the $2^n \times 2^n$ matrix GG^T has coefficient at position $(u,v) \in \{0,1\}^n \times \{0,1\}^n$ equal to

$$\frac{2^{-n}}{Q} \sum_{q=1}^{Q} (-1)^{(u \oplus v) \cdot (t_q \oplus k)} = 2^{-n} \sum_{t \in \mathbb{F}_2^n} \left(\frac{1}{Q} \sum_{q/t_q=t} 1 \right) (-1)^{(u \oplus v) \cdot (t \oplus k)} \xrightarrow[Q \to +\infty]{} \frac{1}{2^n} I_{u,v},$$

by the law of large numbers, where $I_{u,v}$ is the element at position (u,v) in the identity matrix. The limit comes from the fact that $\frac{1}{Q} \sum_{q/t_q=t} 1 \approx \frac{1}{2^n}$ when $Q \to +\infty$, hence the limit using Proposition 7 of [4]. Therefore GG^T is inversed trivially. $\square$

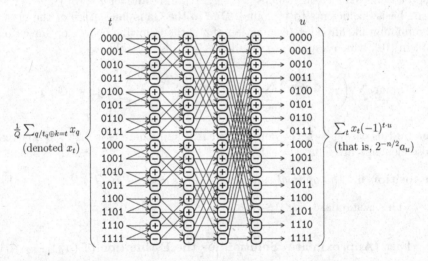

$\frac{1}{Q} \sum_{q/t_q \oplus k=t} x_q$ (denoted x_t)

$\sum_t x_t(-1)^{t \cdot u}$ (that is, $2^{-n/2} a_u$)

Fig. 3. Butterfly algorithm to compute a_u from the average $\frac{1}{Q} \sum_{q/t_q=t} x_q$ using (10)

The expression of a_u given in Eq. (10) is (proportional to) the *(inverse) Fourier transform* of the average of leakage traces in each class $(x_q)_{q/q_t=t}$. It is easily computed as follows:

1. sum the traces per value of t, which yields the vector $(\sum_{q/t_q=t} x_q)_{t \in \mathbb{F}_2^n}$,
2. multiply this vector by the Walsh-Hadamard matrix $\frac{2^{n/2}}{Q} H(2^n)$.

The second step can be optimized with the classical *butterfly* FFT algorithm, which is sketched in Fig. 3 for $n = 4$. Overall, the complexity of the computation of $(a_u)_{u \in \mathbb{F}_2^n}$ from the pairs $(x_q, t_q)_{1 \le q \le Q}$ is $\mathcal{O}(Q + n \cdot 2^n)$.

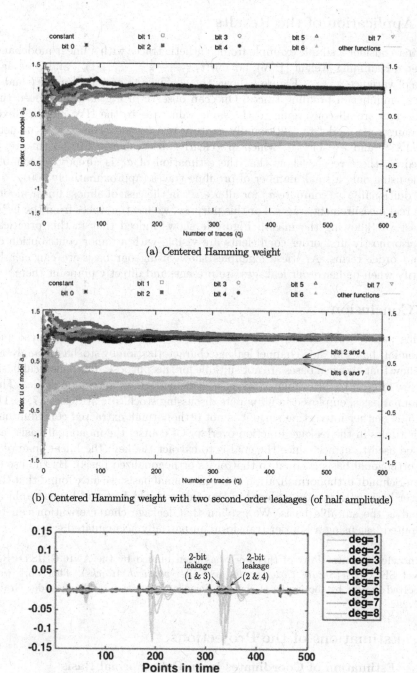

(a) Centered Hamming weight

(b) Centered Hamming weight with two second-order leakages (of half amplitude)

(c) Dpacontest v4 traces (unprotected scenario) [degree 0 is not represented]

Fig. 4. Estimation of coefficients a_u using Fourier transform

5   Application of the Results

We first consider a simple example from synthetic traces with a linear model and centered Hamming Weight (HW), i.e. $w_H(t) = \frac{n}{2} - \frac{1}{2}\sum_{i=0}^{n-1}(-1)^{t_i}$, and Gaussian noise of variance $\sigma^2 = 2$. Figure 4 shows the coefficients a_u^2 for all $u \in \mathbb{F}_2^n$ and a varying number of profiling traces. One can observe in Fig. 4a that indeed the coefficients are all converging to the same value due to the HW model. Next, we change our model to additionally capture two second order terms, namely $\frac{1}{4}(-1)^{t_2+t_4}$ and $\frac{1}{4}(-1)^{t_6+t_7}$, which are clearly observable in Fig. 4b (in grey). Moreover, these results show that the estimation of a_u is already reasonable stable using only a small number of profiling traces (approximatively 200).

Additionally, we compute a_u^2 for all $u \in \mathbb{F}_2^n$ in the case of almost linear model from real measurement traces. For this purpose, we use the traces from the DPA contest v4 (knowing the mask). Figure 4c shows indeed that in this practical scenario mostly first order coefficients are visible with a minor contribution of second order terms. As these examples show, using our basis we can clearly identify when higher order leakages are present, and directly pinpoint them.

6   Conclusion

In this paper, we have discussed the suitability of "classical" (canonical and monomial) bases for side-channel leakage characterization by stochastic analysis. We show that classical bases are not suitable for this purpose: The canonical basis is of few interest to the evaluator because all elements have maximum degree. The monomial basis, employed in all papers discussing stochastic attacks [6,7,10,11, 14,15] is neither interesting since it is not orthonormal: extracted contributions of bit tuples in the leakage function overlap. Of course, the monomial basis can still be used to attack, since the goal is to extract the key (the linear span of a non-orthogonal basis is equal to that of its orthogonalized basis). By the use of Gram-Schmidt orthonormalization of the monomial basis, we have found that the Fourier basis with vectors ordered in Hamming weight first and lexicographical second is the suitable basis. We explain that leakage characterization can be computed fast using a Fourier transform on partially accumulated traces.

Acknowledgments. Part of this work has been funded by the ANR CHIST-ERA project SECODE (*Secure Codes to thwart Cyber-physical Attacks*). This work was supported in part by the National Natural Science Foundation of China under Grant 61472292.

A   Estimations of the Projections

A.1   Estimation of Coordinates in an Orthonormal Basis

We consider a profiling situation where the attacker knows the secret key k, but does not know the model f in Eq. (1). Thanks to an orthonormal basis $(\psi_u)_{u \in \mathbb{F}_2^n}$, the model f can be profiled easily from $(x_q)_{1 \le q \le Q}$ measurements, corresponding to $(t_q)_{1 \le q \le Q}$ (uniformly distributed) plaintexts.

Lemma 8. *Decompose the unknown function f as $f = \sum_{u \in \mathbb{F}_2^n} a_u \psi_u$, where $a_u = \langle f | \psi_u \rangle$. For every $u \in \mathbb{F}_2^n$, a_u is consistently estimated as $\widehat{a_u}$, the empirical correlation[1] between X and $\psi_u(T \oplus k)$:*

$$\widehat{a_u} = \frac{2^n}{Q} \sum_{q=1}^{Q} x_q \psi_u(t_q \oplus k).$$

Proof. By the law of large numbers,

$$\frac{1}{Q} \sum_{q=1}^{Q} x_q \psi_u(t_q \oplus k) \xrightarrow[Q \to +\infty]{} \mathbb{E}(X \psi_u(T \oplus k)).$$

But from Eq. (1),

$$
\begin{aligned}
\mathbb{E}\left(X \psi_u(T \oplus k)\right) &= \mathbb{E}\left((f(T \oplus k) + N)\psi_u(T \oplus k)\right) \qquad (11) \\
&= \mathbb{E}(f(T \oplus k)\psi_u(T \oplus k)) + \underbrace{\mathbb{E}(N\psi_u(T \oplus k))}_{0} \\
&= \mathbb{E}(f(T \oplus k)\psi_u(T \oplus k)) \\
&= \frac{1}{2^n} \sum_{t \in \mathbb{F}_2^n} f(t)\psi_u(t) = \frac{1}{2^n}\langle f | \psi_u \rangle = \frac{1}{2^n} a_u,
\end{aligned}
$$

where the noise term disappeared because N is centered and independent from T, and where the first expectation term is a balanced sum over t because T is uniformly distributed. □

This theoretical result justifies rigorously why it is customary in the side-channel literature to make use of correlation (or the sibling *covariance* tool) to profile a leakage model [3].

A.2   Incorrect Estimation of Coordinates in a Nonorthogonal Basis

We illustrate in the following example why the monomial basis (though extensively used in the side-channel literature [11,14,15]) is not appropriate for estimating the deterministic part (that is, the function f in Eq. (1)) of the leakage model.

Example 9. Let a leakage function $f : \mathbb{F}_2^n \to \mathbb{R}$, which simply consists in $f(t) = t_0 t_1$. In the understanding of the state-of-the-art, this function models the sole interaction of bits 0 and 1 of bitvector $t = (t_i)_{0 \le i \le n-1}$.

We show that the blind application of the above correlation method (Lemma 8) does not allow to recover easily the fact that f consists in the interaction between bits 0 and 1. In fact, letting $u \in \mathbb{F}_2^n$, the correlation between the monomial basis vector ϕ_u and leakage X (Eq. (11)) equals

---

[1] The term *correlation* is used here in the sense of *scalar product* between two data series. This shall not be confused with the *Pearson correlation coefficient* used, for instance, in the *Correlation Power Analysis* [2].

$$a_u = 2^n \mathbb{E}(X \phi_u(T \oplus k)) \tag{12}$$

$$= \sum_{t \in \mathbb{F}_2^n} t_0 t_1 \, \phi_u(t) \qquad \text{(by the change of variable } t \leftarrow t \oplus k\text{)}$$

$$= \sum_{t \in \mathbb{F}_2^n} t_0 t_1 \prod_{i/u_i=1} t_i = \sum_{t \in \mathbb{F}_2^n} \prod_{i \in \{0,1\} \cup \{i/u_i=1\}} t_i = 2^{n-2-\sum_{i=2}^{n-1} u_i}$$

$$= \begin{cases} 2^{n-2} & \text{for } u = (0,0,0,\dots,0),(1,0,0,\dots,0),(0,1,0,\dots,0),(1,1,0,\dots,0); \\ 2^{n-3} & \text{for all } u \text{ such that } \sum_{i=2}^{n-1} u_i = 1, e.g., u = (0,0,0,\dots,0,1), \\ & \quad (1,0,0,\dots,0,1),(0,1,0,\dots,0,1),(1,1,0,\dots,0,1), \text{ etc.} \\ \vdots & \\ 2 & \text{for } u \text{ such that } \sum_{i=2}^{n-1} u_i = n-3, \text{and} \\ 1 & \text{for } u \text{ such that } \sum_{i=2}^{n-1} u_i = n-2. \end{cases} \tag{13}$$

While the value of a_u is indeed largest for $u = (1,1,0,\dots,0)$ as expected, this maximum value ($=2^{n-2}$) is also reached by $u = (1,0,0,\dots,0)$ and $=(0,1,0,\dots,0)$, which represent single bits. Moreover, there are non-zero terms (albeit smaller) for coefficients a_u such that $w_H(u) > 2$.

Therefore, the covariance method is clearly ill-fitted to characterize that particular leakage function f. The reason for this failure is of course that Lemma 8 is applied in this (counter-)example using the monomial basis $(\phi_u)_{u \in \mathbb{F}_2^n}$, which is not orthonormal.

In summary, we face the problem that the leakage model f cannot be characterized using the *covariance* tool in the monomial basis. This explains why, from Sect. 3 onwards, we investigate a suitable basis, which should have both properties of: (1) being orthonormal (for easy application of the covariance method of Lemma 8) and (2) being interpretable in terms of bits interaction. This will allow to select which vectors of the basis to keep when performing an attack.

References

1. Banerjee, S., Roy, A.: Linear Algebra and Matrix Analysis for Statistics. Texts in Statistical Science, 1st edn. Chapman and Hall/CRC, Hoboken (2014). ISBN 978-1420095388

2. Brier, E., Clavier, C., Olivier, F.: Correlation power analysis with a leakage model. In: Joye, M., Quisquater, J.-J. (eds.) CHES 2004. LNCS, vol. 3156, pp. 16–29. Springer, Heidelberg (2004). doi:10.1007/978-3-540-28632-5_2

3. Bruneau, N., Danger, J.-L., Guilley, S., Heuser, A., Teglia, Y.: Boosting higher-order correlation attacks by dimensionality reduction. In: Chakraborty, R.S., Matyas, V., Schaumont, P. (eds.) SPACE 2014. LNCS, vol. 8804, pp. 183–200. Springer, Cham (2014). doi:10.1007/978-3-319-12060-7_13

4. Carlet, C.: Boolean functions for cryptography and error correcting codes. In: Crama, Y., Hammer, P. (eds.) Chapter of the Monography Boolean Models and Methods in Mathematics, Computer Science, and Engineering, pp. 257–397. Cambridge University Press (2010)

5. Chari, S., Rao, J.R., Rohatgi, P.: Template attacks. In: Kaliski, B.S., Koç, K., Paar, C. (eds.) CHES 2002. LNCS, vol. 2523, pp. 13–28. Springer, Heidelberg (2003). doi:10.1007/3-540-36400-5_3
6. Gierlichs, B., Lemke-Rust, K., Paar, C.: Templates vs. stochastic methods. In: Goubin, L., Matsui, M. (eds.) CHES 2006. LNCS, vol. 4249, pp. 15–29. Springer, Heidelberg (2006). doi:10.1007/11894063_2
7. Heuser, A., Kasper, M., Schindler, W., Stöttinger, M.: How a symmetry metric assists side-channel evaluation - a novel model verification method for power analysis. In: Proceedings of the 14th Euromicro Conference on Digital System Design (DSD 2011), Washington, DC, pp. 674–681. IEEE Computer Society (2011)
8. Heuser, A., Kasper, M., Schindler, W., Stöttinger, M.: A new difference method for side-channel analysis with high-dimensional leakage models. In: Dunkelman, O. (ed.) CT-RSA 2012. LNCS, vol. 7178, pp. 365–382. Springer, Heidelberg (2012). doi:10.1007/978-3-642-27954-6_23
9. Heuser, A., Rioul, O., Guilley, S.: Good is not good enough. In: Batina, L., Robshaw, M. (eds.) CHES 2014. LNCS, vol. 8731, pp. 55–74. Springer, Heidelberg (2014). doi:10.1007/978-3-662-44709-3_4
10. Heuser, A., Schindler, W., Stöttinger, M.: Revealing side-channel issues of complex circuits by enhanced leakage models. In: Rosenstiel, W., Thiele, L. (eds.) DATE, pp. 1179–1184. IEEE (2012)
11. Kasper, M., Schindler, W., Stöttinger, M.: A stochastic method for security evaluation of cryptographic FPGA implementations. In: Bian, J., Zhou, Q., Athanas, P., Ha, Y., Zhao, K. (eds.) FPT, pp. 146–153. IEEE (2010)
12. Lomné, V., Prouff, E., Roche, T.: Behind the scene of side channel attacks. In: Sako, K., Sarkar, P. (eds.) ASIACRYPT 2013. LNCS, vol. 8269, pp. 506–525. Springer, Heidelberg (2013). doi:10.1007/978-3-642-42033-7_26
13. Renauld, M., Standaert, F.-X., Veyrat-Charvillon, N., Kamel, D., Flandre, D.: A formal study of power variability issues and side-channel attacks for nanoscale devices. In: Paterson, K.G. (ed.) EUROCRYPT 2011. LNCS, vol. 6632, pp. 109–128. Springer, Heidelberg (2011). doi:10.1007/978-3-642-20465-4_8
14. Schindler, W.: On the optimization of side-channel attacks by advanced stochastic methods. In: Vaudenay, S. (ed.) PKC 2005. LNCS, vol. 3386, pp. 85–103. Springer, Heidelberg (2005). doi:10.1007/978-3-540-30580-4_7
15. Schindler, W., Lemke, K., Paar, C.: A stochastic model for differential side channel cryptanalysis. In: Rao, J.R., Sunar, B. (eds.) CHES 2005. LNCS, vol. 3659, pp. 30–46. Springer, Heidelberg (2005). doi:10.1007/11545262_3
16. Standaert, F.-X., Koeune, F., Schindler, W.: How to compare profiled side-channel attacks? In: Abdalla, M., Pointcheval, D., Fouque, P.-A., Vergnaud, D. (eds.) ACNS 2009. LNCS, vol. 5536, pp. 485–498. Springer, Heidelberg (2009). doi:10.1007/978-3-642-01957-9_30
17. Zhao, H., Zhou, Y., Standaert, F.-X., Zhang, H.: Systematic construction and comprehensive evaluation of kolmogorov-smirnov test based side-channel distinguishers. In: Deng, R.H., Feng, T. (eds.) ISPEC 2013. LNCS, vol. 7863, pp. 336–352. Springer, Heidelberg (2013). doi:10.1007/978-3-642-38033-4_24

Key-Policy Attribute-Based Encryption
from Bilinear Maps

Ferucio Laurențiu Țiplea[1]([⊠]), Constantin Cătălin Drăgan[2],
and Anca-Maria Nica[1]

[1] Department of Computer Science, "Alexandru Ioan Cuza" University of Iași,
700506 Iași, Romania
ferucio.tiplea@uaic.ro, nica.anca@student.uaic.ro
[2] CNRS, LORIA, 54506 Vandoeuvre-lès-Nancy Cedex, France
catalin.dragan@loria.fr

Abstract. The aim of this paper is to provide an overview on the
newest results regarding the design of key-policy attribute-based encryption (KP-ABE) schemes from secret sharing and bilinear maps.

1 Introduction

Attribute-based encryption (ABE) is a new paradigm in cryptography, where
messages are encrypted and decryption keys are computed in accordance with
a given set of attributes and an access structure on the set of attributes. There
are two forms of ABE: *key-policy ABE* (KP-ABE) [8] and *ciphertext-policy ABE*
(CP-ABE) [2]. In a KP-ABE, each message is encrypted together with a set of
attributes and the decryption key is computed for the entire access structure; in
a CP-ABE, each message is encrypted together with an access structure while
the decryption keys are given for specific sets of attributes.

In this paper we focus only on KP-ABE. The first KP-ABE scheme was
proposed in [8], where the access structures were specified by monotone Boolean
formulas (monotone Boolean circuits of fan-out one, with one output wire). An
extension to the non-monotonic case has later appeared in [10]. Both approaches
[8,10] take into consideration only access structures defined by Boolean formulas.
However, there are access structures of practical importance that cannot be
represented by Boolean formulas, such as multi-level access structures [13,14].
In such a case, defining KP-ABE schemes to work with general Boolean circuits
becomes a necessity. The first solution to this problem was proposed in [6] by
using leveled multi-linear maps. A little later, a lattice-based construction was
also proposed [7].

Several construction of KP-ABE schemes based on bilinear maps were proposed. The first one proposed in [8] works in two steps: in the first step, a secret
is top-down shared on a Boolean tree, while in the second step some information is bottom-up reconstructed using just one bilinear map. The scheme is very
appealing and practically efficient. However, it works only with Boolean trees

© Springer International Publishing AG 2017
P. Farshim and E. Simion (Eds.): SecITC 2017, LNCS 10543, pp. 28–42, 2017.
https://doi.org/10.1007/978-3-319-69284-5_3

(formulas); a direct extension of it to general Boolean circuits faces the back-tracking attack [6]. The second construction [6] works in just one step which is a bottom-up reconstruction of some information, by means of a leveled multi-linear map (sequence of bilinear maps with special constraints). The scheme can be used with general Boolean circuits but is much less efficient than the one in [8]: the decryption key size depends on the number of gates of the Boolean circuit and the leveled multi-linear maps are more complex structures than bilinear maps. Moreover, leveled multi-linear maps of some depth k do not easily scale to fit Boolean circuits of depth larger than $k + 1$.

Whether KP-ABE schemes for general Boolean circuits can be constructed using only bilinear maps, is still an open question. An attempt to solve this problem would be to look for methods of top-down secret sharing on Boolean circuits, capable to defeat the backtracking attack. Three such methods were recently proposed. The first one [3] extends the scheme in [8] to work with general Boolean circuits. The scheme is practically efficient only for a subclass of Boolean circuits which strictly extends the class of Boolean formulas (and, therefore, it is a proper extension of the scheme in [8]). The second method [4], when used in conjunction with simplified forms of leveled multi-linear maps, gives rise to a scheme which works for general Boolean circuits and is much efficient than the scheme in [6]. The thirdmethod [9] is a slight refinement of the one in [3], resulting in shorter decryption keys. All these schemes are secure in the selective model.

2   Attribute-Based Encryption and the Backtracking Attack

We recall below a few concepts and notations on attribute-based encryption; for details the reader is referred to [6,8] which are the main papers we build on.

Access structures. It is customary to represent access structures [12] by Boolean circuits [1]. A Boolean circuit has a number of input wires (which are not gate output wires), a number of output wires (which are not gate input wires), and a number of OR-, AND-, and NOT-gates. The OR- and AND-gates have a fan-in of two, while NOT-gates have a fan-in of one. All of them have a fan-out of at least one. Boolean circuits where all gates have a fan-out of one correspond to *Boolean formulas*. A Boolean circuit is *monotone* if it does not have NOT-gates. In this paper all Boolean circuits have exactly one output wire and are monotone (the restriction to monotone Boolean circuits does not constitute a loss of generality, as it was pointed out in [6]).

If the input wires of a Boolean circuit $\mathcal{C}$ are in a one-to-one correspondence with the elements of a set $\mathcal{U}$ of elements called *attributes*, we will say that $\mathcal{C}$ is a Boolean circuit over $\mathcal{U}$. Each $A \subseteq \mathcal{U}$ evaluates the circuit $\mathcal{C}$ to one of the Boolean values 0 or 1 by simply assigning 1 to all input wires associated to elements in A, and 0 otherwise; then the Boolean values are propagated bottom-up to all gate output wires in a standard way. $\mathcal{C}(A)$ stands for the Boolean value obtained by evaluating $\mathcal{C}$ for A. The access structure defined by $\mathcal{C}$ is the set of all A with $\mathcal{C}(A) = 1$.

Attribute-based encryption. A KP-ABE scheme consists of four probabilistic polynomial-time (PPT) algorithms [8]:

$Setup(\lambda)$: this is a PPT algorithm that takes as input the security parameter λ and outputs a set of public parameters PP and a master key MSK;

$Enc(m, A, PP)$: this is a PPT algorithm that takes as input a message m, a non-empty set of attributes $A \subseteq \mathcal{U}$, and the public parameters, and outputs a ciphertext E;

$KeyGen(\mathcal{C}, MSK)$: this is a PPT algorithm that takes as input a Boolean circuit $\mathcal{C}$ and the master key MSK, and outputs a decryption key D;

$Dec(E, D)$: this is a deterministic polynomial-time algorithm that takes as input a ciphertext E and a decryption key D, and outputs a message m or the special symbol $\perp$.

The following correctness property is required to be satisfied by any KP-ABE scheme: for any $(PP, MSK) \leftarrow Setup(\lambda)$, any Boolean circuit $\mathcal{C}$ over a set $\mathcal{U}$ of attributes, any message m, any $A \subseteq \mathcal{U}$, and any $E \leftarrow Enc(m, A, PP)$, if $\mathcal{C}(A) = 1$ then $m = Dec(E, D)$, for any $D \leftarrow KeyGen(\mathcal{C}, MSK)$.

The first KP-ABE scheme was proposed in [8]. It work only for Boolean formulas and uses secret sharing and just one bilinear map. Its main idea is as follows:

- let $e : G_1 \times G_1 \rightarrow G_2$ be a bilinear map and g a generator of G_1, where G_1 and G_2 are groups of prime order p (recall that e is bilinear if $e(u^a, v^b) = e(u, v)^{ab}$ for all $u, v \in G_1$ and $a, b \in \mathbb{Z}_p$, and $e(g, g)$ is a generator of G_2, provided that g is a generator of G_1);
- to encrypt a message $m \in G_2$ by a set A of attributes, just multiply m by $e(g, g)^{ys}$, where y is a random integer chosen in the setup phase and s is a random integer chosen in the encryption phase. Moreover, an attribute dependent quantity is also computed for each attribute $i \in A$;
- the decryption key is generated by sharing y top-down on the Boolean circuit; then, key components are associated to the input wires, based on the corresponding shares;
- in order to decrypt $me(g, g)^{ys}$ by a set A of attributes, one has to compute $e(g, g)^{ys}$. This can be done only if A is authorized and, in such a case, the computation of $e(g, g)^{ys}$ is bottom-up, starting from the key components associated to the input wires.

It was pointed out in [6] that the construction in [8] cannot be used with general Boolean circuits. The reason is the next one. Due to the way the secret y is shared, both input wires of OR-gates get the value in the computation process of $e(g, g)^{ys}$ (see the notations above). This shows that the value computed at an OR-gate by means of one of its input wires may freely migrate back-down to the other input wire of the gate, and then to other wires if some gates are of fan-out more than one (an illustration of this attack can be found in [3]). This attack, called the *backtracking attack*, cannot occur in case of Boolean formulas because, in such a case, the input wires of OR-gates are not used by any other gates.

To avoid the backtracking attack, [6] renounces to the secret sharing phase and uses a "one-way" construction in evaluating the Boolean circuit. The idea is the next one:

- consider a *leveled multi-linear map*, which consists of k cyclic groups $G_1, \ldots, G_k$ of prime order p, k generators $g_1, \ldots, g_k$ of these groups, respectively, and a set $\{e_{i,j} : G_i \times G_j \to G_{i+j} | i, j \geq 1, \ i + j \leq k\}$ of bilinear maps satisfying $e_{i,j}(g_i^a, g_j^b) = g_{i+j}^{ab}$, for all i and j and all $a, b \in \mathbb{Z}_p^*$, where k is the circuit depth plus one;
- the key components are associated to each input wire and to each gate output wire;
- the circuit is evaluated bottom-up and the values on level j are powers of g_{j+1};
- as the mappings $e_{i,j}$ work only in the "forward" direction, it is not feasible to invert values on the level $j + 1$ to obtain values on the level j, defeating thus the backtracking attack.

As with respect to the existence of leveled multi-linear maps, [6] shows how this scheme can be translated into the GGH graded algebra framework [5].

Conventions on Boolean circuits. Assuming that the wires of Boolean circuits are labeled, we refer to logic gates as tuples $(w_1, w_2, X, w_3, \ldots, w_j)$ for some $j \geq 3$ and $X \in \{OR, AND\}$. The elements before (after) the gate name are the input (output) wires of the gate. The output wire of a Boolean circuit will always be denoted by o, and the input wires by $1, \ldots, n$ (assuming that the circuit has n input wires).

To clearly understand the secret sharing procedures we propose, we decompose the input and logic gates of fan-out more than one into two gates: a gate of fan-out one and a special *fanout-gate* (FO-gate) which multiplies the output of the former gate. For instance, an AND-logic gate $(w_1, w_2, AND, w_3, w_4)$ will be decomposed into (w_1, w_2, AND, w) and (w, FO, w_3, w_4) (see Figs. 1, 2, and 3 for examples).

The gates of a Boolean circuit $\mathcal{C}$ are distributed on levels, starting with the level 0 which consists of the input gates. A FO-gate on top of an input or logic gate Γ is considered to be on the same level as Γ (it is customary to assume that no two FO-gates are directly connected). The depth of $\mathcal{C}$, denoted $depth(\mathcal{C})$, is the number of the output level of $\mathcal{C}$ (see Fig. 2). A level which contains FO-gates is called a *FO-level* (for instance, the level 1 in Figs. 1 and 3, and the level 0 in Fig. 2). Given a logic gate Γ on some level i with its left input gate Γ' on some level $j < i$, the *left FO-level sequence* of Γ is the sequence of FO-level numbers taken in decreasing order from $i - 1$ to $j + 1$, if Γ' is an FO-gate, and to j, otherwise. In a similar way are defined the *right FO-level sequences*. Clearly, these sequences may be empty. For instance, (1) is the left FO-level sequence of Γ_2, and the right FO-level sequence of Γ_3, in the Boolean circuits in Fig. 1. The empty sequence is a right FO-level sequence of Γ_2, and a left FO-level sequence of Γ_3.

3   ABE from Secret Sharing on Boolean Circuits and Bilinear Maps

In this section we present the KP-ABE scheme proposed in [3]. The scheme defeats the backtracking attack and is practically efficient for a proper extension of the Boolean formulas (Subsect. 3.2 provides details). Therefore, it can be regarded as a proper extension of the scheme in [8]. Due to the space limitations, some of the technical details and proofs are omitted but they can be found in [3].

3.1   KP-ABE_Scheme_1 and Its Security

As explained in Sect. 1, a natural idea of extending the scheme in [8] to Boolean circuits is to look for another secret sharing procedure capable to share top-down a secret y on a Boolean circuit $\mathcal{C}$ so that the backtracking attack is defeated. To this, the sharing procedure $Share(y, \mathcal{C})$ below is used, which outputs two functions S and P with the following meaning: S assigns to each wire of $\mathcal{C}$ a list of values in $\mathbb{Z}_p$, while P assigns to each output wire of a FO-gate a list of pairs of values in G_1. The notations used to describe $Share(y, \mathcal{C})$ are exactly those in Sect. 2 used to describe the scheme in [8]. Supplementary, a few concepts on lists are in order. A *list of length* n over a set X is any vector $L \in X^n$. $|L|$ stands for the length of L, $L_1 L_2$ for the *concatenation* of two lists L_1 and L_2, $pos(L) = \{1, \ldots, |L|\}$, and $L(i)$ denotes the ith element of L. If L is a list of lists, then $L(i, j)$ denotes the jth element of the list $L(i)$.

$Share(y, \mathcal{C})$

1. Initially, all gates of $\mathcal{C}$ are unmarked and $S(o) := (y)$;
2. If $\Gamma = (w_1, w_2, OR, w)$ is unmarked, then:
   - mark it,
   - $S(w_1) := S(w)$, and
   - $S(w_2) := S(w)$;
3. If $\Gamma = (w_1, w_2, AND, w)$ is unmarked, then:
   - mark it,
   - $S(w_1) := (x_i^1 | 1 \leq i \leq |S(w)|)$, and
   - $S(w_2) := (x_i^2 | 1 \leq i \leq |S(w)|)$,
   where $x_i^1 \leftarrow \mathbb{Z}_p$ and x_i^2 satisfies $S(w)(i) = (x_i^1 + x_i^2) \bmod p$, for each $i \in pos(S(w))$;
4. If $\Gamma = (w, FO, w_1, \ldots, w_j)$ is unmarked, then:
   - mark it,
   - $S(w) := S(w_1)' \cdots S(w_j)'$, and
   - $P(w_k) := (g^{b_i} | 1 \leq i \leq |S(w_k)|)$,
   where $S(w_k)' = (a_i | 1 \leq i \leq |S(w_k)|)$, $a_i \leftarrow \mathbb{Z}_p$ and b_i satisfies $S(w_k)(i) = (a_i + b_i) \bmod p$, for all $1 \leq k \leq j$ and $i \in pos(S(w_k))$;
5. repeat the last three steps above until all gates get marked.

Figure 1 illustrates the procedure *Share*.

The deterministic procedure $Recon(\mathcal{C}, P, V, g^s)$ reconstructs $e(g,g)^{ys}$ from the shares associated to some set A of attributes. In fact, it outputs an evaluation function R which assigns to each wire a list of values in $G_2 \cup \{\bot\}$. The notation and conventions here are as follows:

- $(S, P) \leftarrow Share(y, \mathcal{C})$, for some $y \in \mathbb{Z}_p$ and Boolean circuit $\mathcal{C}$ with n input wires;
- $s \in \mathbb{Z}_p$;
- $V = (V(i) | 1 \leq i \leq n)$, where $V(i)$ is either a list $(e(g,g)^{\alpha_i} | 1 \leq j \leq |S(i)|)$ for some $\alpha_i \in \mathbb{Z}_p$, or a list of $|S(i)|$ undefined values $\bot$, for all $1 \leq i \leq n$

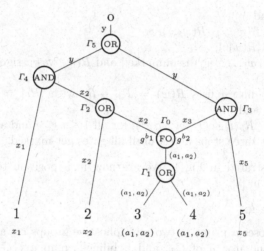

$$x_1 + x_2 \equiv y \bmod p, \; x_3 + x_5 \equiv y \bmod p, \; a_1 + b_1 \equiv x_2 \bmod p, \; a_2 + b_2 \equiv x_3$$

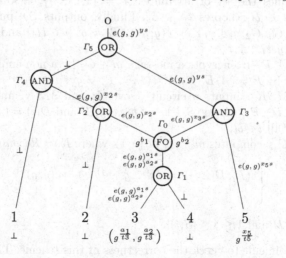

Fig. 1. *Share(y, $\mathcal{C}$)* (on top); *Recon($\mathcal{C}, P, V, g^s$)* (on bottom)

($\perp$ satisfies $\perp < x$ for all $x \in G_2$, and $\perp \cdot z = \perp$, $z/\perp = \perp$, and $\perp^z = \perp$, for all $z \in G_2 \cup \{\perp\}$).

$\underline{Recon(\mathcal{C}, P, V, g^s)}$

1. Initially, all gates of $\mathcal{C}$ are unmarked and $R(i) := V(i)$, for all $i \in \mathcal{U}$;
2. If $\Gamma = (w_1, w_2, OR, w)$ is unmarked and both $R(w_1)$ and $R(w_2)$ were assigned, then:
   – mark Γ and
   – $R(w, i) := sup\{R(w_1, i), R(w_2, i)\}$,
   for all $1 \le i \le |R(w_1)|$;
3. If $\Gamma = (w_1, w_2, AND, w)$ is unmarked and both $R(w_1)$ and $R(w_2)$ were assigned, then:
   – mark Γ and
   – $R(w, i) := R(w_1, i) \cdot R(w_2, i)$,
   for all $1 \le i \le |R(w_1)|$;
4. If $\Gamma = (w, FO, w_1, \ldots, w_j)$ is unmarked and $R(w)$ was assigned, then:
   – mark Γ,
   – split $R(w)$ into j lists $R(w) = R_1 \cdots R_j$ with $|R_k| = |P(w_k)|$ for all $1 \le k \le j$, and
   – $R(w_k, i) := R_k(i) \cdot e(P(w_k, i), g^s)$, for all $1 \le k \le j$ and all $1 \le i \le |R_k|$;
5. repeat the last three steps above until all gates get marked.

Recon is illustrated in Fig. 1. We are now in a position to introduce the scheme proposed in [3].

$\underline{KP\text{-}ABE\_Scheme\_1}$

Setup(λ, n): chooses a prime p, two multiplicative groups G_1 and G_2 of prime order p, a generator g of G_1, and a bilinear map $e. : G_1 \times G_1 \to G_2$. Then, defines the set of attributes $\mathcal{U} = \{1, \ldots, n\}$, chooses $y \in \mathbb{Z}_p$ and, for each $i \in \mathcal{U}$, chooses $t_i \in \mathbb{Z}_p$. Finally, outputs the public parameters $PP = (p, G_1, G_2, g, e, n, Y = e(g, g)^y, (T_i = g^{t_i} | i \in \mathcal{U}))$ and the master key $MSK = (y, t_1, \ldots, t_n)$;

Encrypt(m, A, PP): encrypts a message $m \in G_2$ by a non-empty set $A \subseteq \mathcal{U}$ of attributes by $E = (A, E' = mY^s, (E_i = T_i^s = g^{t_i s} | i \in A), g^s)$, where $s \leftarrow \mathbb{Z}_p$;

KeyGen$(\mathcal{C}, MSK)$: inputs a circuit $\mathcal{C}$ over $\mathcal{U}$ and MSK, and outputs $D = ((D(i) | i \in \mathcal{U}), P)$, where $(S, P) \leftarrow Share(y, \mathcal{C})$ and $D(i) = (g^{S(i,j)/t_i} | 1 \le j \le |S(i)|)$ for all $i \in \mathcal{U}$;

Decrypt(E, D): computes $m := E'/R(o, 1)$, where $R := Recon(\mathcal{C}, P, V_A, g^s)$ and

$$V_A(i, j) = \begin{cases} e(E_i, D(i, j)) = e(g^{t_i s}, g^{S(i,j)/t_i}) = e(g, g)^{S(i,j)s}, & \text{if } i \in A \\ \perp, & \text{otherwise} \end{cases}$$

for all $i \in \mathcal{U}$ and $1 \le j \le |S(i)|$.

It is not difficult to check the correctness of this scheme. The next theorem states its security (its proof can be found in [3]).

Theorem 1. *The KP-ABE_Scheme_1 is secure in the selective model under the decisional bilinear Diffie-Hellman assumption.*

Complexity. Assume that the Boolean circuit has n input wires and r FO-gates of fan-out at most j. If there is no path between any two FO-gates, by the sharing procedure, exactly r input wires will receive at most j shares (but at least two), and the other input wires will receive exactly one share. This leads to at most $n + r(j - 1)$ key components, and this is the minimum size of the decryption key (remark also that $r \leq n$ in this case). If there are paths between FO-gates, then the maximum number of shares some input wires of the circuit may receive is at most j^α, where α is the number of levels that contain FO-gates (α is less than or equal to minimum of r and the circuit depth).

3.2 Applications and Comparisons

The above sub-section shows clearly that KP-ABE_Scheme_1 is practically inefficient if many FO-gates are path-connected. However, if the FO-gates are on the lowest levels and not too much path-connected, the scheme may be practically efficient (and more efficient than the one in [6]). We will illustrate this on multi-level access structures [11,13]. A *disjunctive multi-level access structure* [11] over a set $\mathcal{U}$ of attributes is a tuple $(\overline{a}, \overline{\mathcal{U}}, \mathcal{S})$, where $\overline{a} = (a_1, \ldots, a_k)$ is a vector of positive integers satisfying $0 < a_1 < \cdots < a_k$, $\overline{\mathcal{U}} = (\mathcal{U}_1, \ldots, \mathcal{U}_k)$ is a partition of $\mathcal{U}$, and $\mathcal{S} = \{A \subseteq \mathcal{U} | (\exists 1 \leq i \leq k)(|A \cap (\cup_{j=1}^{i} \mathcal{U}_j)| \geq a_i)\}$. If we replace "$\exists$" by "$\forall$" in the definition of $\mathcal{S}$, we obtain *conjunctive multi-level access structures* [13].

It is well-known, and not difficult to prove [3], that disjunctive and conjunctive multi-level access structures cannot be represented by Boolean formulas. However, they can be represented by Boolean circuits. For a compact representation of them we endow the Boolean circuits by (a, b)-*threshold gate* [8], where $1 \leq a \leq b$ and $b \geq 2$, which are logical gates with b input wires and one output wire. The output wire of such a gate is evaluated to the truth value 1 whenever at least a input wires of the gate are assigned to the truth value 1. OR-gates are $(1, 2)$-threshold gates, while AND-gates are $(2, 2)$-threshold gates. With such Boolean circuits, multi-level access structures can compactly be represented as in Fig. 2.

KP-ABE_Scheme_1 can easily be adapted to accommodate threshold gates by using a probabilistic linear secret sharing scheme. Then, a simple comparison between our scheme and the one in [6] for multi-level access structures as those above, leads to Table 1 below which shows that our scheme performs much better than the one in [6], where n_1 and $n_i - n_{i-1}$ were approximated by n/k, for all $2 \leq i \leq k$ (full details are provided in [3]).

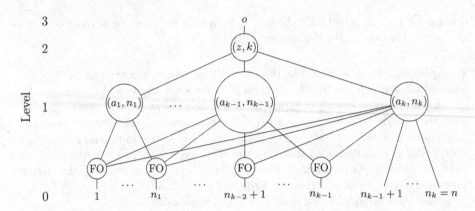

Fig. 2. Boolean circuit representation of multilevel access structure: z is 1 for the disjunctive case, and k for the conjunctive case.

Table 1. Comparisons between the scheme in [6] and KP-ABE_Scheme_1 for multi-level access structures

Scheme	Average no. of keys		No. of bilinear maps
KP-ABE scheme in [6]	Case 1 ($a_i = n_i$, $\forall i$): $n\frac{k+5}{2} + 3k + 1 - z$		3
	Case 2 ($a_i < n_i$, $\forall i$): $\geq \left(2 + \frac{(k+1)(k+5)}{3}\right)n + 2k + 1 - z$		
KP-ABE_Scheme_1	$n\frac{k+1}{2}$		1

4   ABE from Secret Sharing on Boolean Circuits and Multi-linear Maps

The first KP-ABE scheme that works for general Boolean formulas was proposed in [6]. The scheme does not use secret sharing; it processes the Boolean circuit bottom-up and changes the generator with each level by means of the leveled multi-linear map. Each wire, except for the output wire, of the circuit gets two, three, or four keys, depending on the wire.

In this section we show that a combination of secret sharing and leveled multi-linear maps [4] leads to a more efficient scheme than the one in [6]. Due to the space limitations, some technical details and proofs are omitted but they can be found in [4].

4.1   KP-ABE_Scheme_2 and Its Security

This new scheme uses a simpler form of leveled multi-linear maps, called *chained multi-linear maps*. A *chained multi-linear map* is a sequence $(e_i : G_i \times G_1 \to G_{i+1} | 1 \leq i \leq k)$ of bilinear maps, where $G_1, \ldots, G_{k+1}$ are multiplicative groups of the same prime order p. If $g_1 \in G_1$ is a generator of G_1, then g_{i+1} defined recursively by $g_{i+1} = e_i(g_i, g_1)$ is a generator of G_{i+1}, for all $1 \leq i \leq k$ (because e_i is a bilinear map). Therefore, $(e_i | 1 \leq i \leq k)$ can be regarded as a special form of leveled multi-linear map.

Chained multi-linear maps will be used in our construction as follows. Assume that r is the number of FO-levels in the Boolean circuits we consider, and $(e_i | 1 \leq i \leq r + 1)$ is a chained multi-linear map as above. A message $m \in G_{r+2}$ will be encrypted by mg_{r+2}^{ys}, where y is a random integer chosen in the setup phase and s is a random integer chosen in the encryption phase. To decrypt this message, one needs to compute g_{r+2}^{ys}, and this will be done by using a secret sharing procedure and a secret reconstruction procedure.

Unlike our previous KP-ABE scheme, the secret sharing procedure $Share(y, \mathcal{C})$ outputs three functions S, P, and L with the following meaning: S assigns an element in $\mathbb{Z}_p$ to each wire of $\mathcal{C}$, P assigns a *FO-key* to each output wire of a FO-gate, and L assigns a *FO-level-key* to each FO-level (see Sect. 2 on concepts and conventions regarding Boolean circuits).

$Share(y, \mathcal{C})$

1. Initially, all gates of $\mathcal{C}$ are unmarked and $S(o) := y$;
2. $L(i) := g_1^{a_i}$ for each FO-level $0 \leq i < depth(\mathcal{C}) - 2$, where $a_i \leftarrow \mathbb{Z}_p$;
3. If $\Gamma = (w_1, w_2, OR, w)$ is unmarked, then:
   - mark Γ,
   - $S(w_1) := S(w)a_{i_1}^{-1} \cdots a_{i_u}^{-1} \bmod p$, and
   - $S(w_2) := S(w)a_{j_1}^{-1} \cdots a_{j_v}^{-1} \bmod p$,
   where $i_1 \cdots i_u$ and $j_1 \cdots j_v$ are the left and right FO-level sequences of Γ, respectively;
4. If $\Gamma = (w_1, w_2, AND, w)$ is unmarked, then:
   - mark Γ,
   - $S(w_1) := x_1$, and
   - $S(w_2) := x_2$,
   where $x_1 \leftarrow \mathbb{Z}_p$, x_2 satisfies $S(w) = (x_1 a_{i_1} \cdots a_{i_u} + x_2 a_{j_1} \cdots a_{j_v}) \bmod p$, and $i_1 \cdots i_u$ and $j_1 \cdots j_v$ are the left and right FO-level sequences of Γ, respectively;
5. If $\Gamma = (w, FO, w_1, \ldots, w_j)$ is unmarked, then:
   - mark Γ,
   - $S(w) \leftarrow \mathbb{Z}_p$, and
   - $P(w_i) := g_1^{b_i}$,
   where b_i satisfies $S(w_i) = S(w)b_i \bmod p$ for all $1 \leq i \leq j$;
6. repeat the last three steps above until all gates get marked.

We write $(S, P, L) \leftarrow Share(y, \mathcal{C})$ to denote that (S, P, L) is an output of the probabilistic algorithm $Share$ on input $(y, \mathcal{C})$. Figure 3 illustrates the procedure $Share$.

The deterministic procedure $Recon(\mathcal{C}, P, L, A, V_A)$ we define below outputs an evaluation function R which assigns to each wire either a value in some group $G_1, \ldots, G_{r+2}$ or the undefined value $\bot$, where r is the number of FO-levels of $\mathcal{C}$. The notation and conventions here are:

- $(S, P, L) \leftarrow Share(y, \mathcal{C})$, for some secret y and Boolean circuit $\mathcal{C}$ with n input wires;
- $A \subseteq \{1, \ldots, n\}$ is a subset of attributes (input wires of $\mathcal{C}$) and $V_A = (V_A(i)|1 \leq i \leq n)$, where $V_A(i) = g_2^{\alpha_i}$ for all $i \in A$ and some $\alpha_i \in \mathbb{Z}_p$, and $V_A(i) = \perp$ for all $i \notin A$;
- $\perp$ is an *undefined value* for which the following conventions are adopted: $\perp \notin \cup_{i=1}^{r+2} G_i$, $\perp < x$, $\perp \cdot z = \perp$, $z/\perp = \perp$, and $\perp^z = \perp$, for all $x \in \cup_{i=1}^{r+2} G_i$ and $z \in (\cup_{i=1}^{r+2} G_i) \cup \{\perp\}$, where r is the number of FO-levels of $\mathcal{C}$.

Before describing the secret reconstruction procedure, one more notation is needed. Given $g_i^\alpha \in G_i$ for some i and α, an FO-level sequence $i_1 \cdots i_u$, and an output L of the *Share* procedure, denote by $Shift(g_i^\alpha, i_1 \cdots i_u, L)$ the element $g_{i+u}^{\alpha a_{i_1} \cdots a_{i_u}} \in G_{i+u}$ obtained as follows:

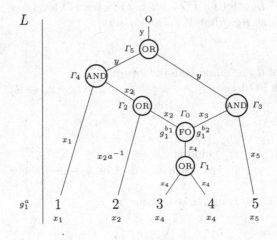

$$x_1 a + x_2 \equiv y \bmod p, \quad x_3 + x_5 a \equiv y \bmod p, \quad x_4 b_1 \equiv x_2 \bmod p, \quad x_4 b_2 \equiv x_3 \bmod p$$

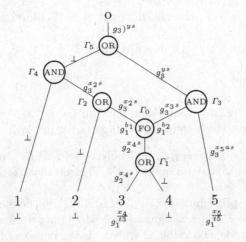

Fig. 3. $Share(y, \mathcal{C})$ (on top); $Recon(\mathcal{C}, P, L, A, V_A)$ with $A = \{3, 5\}$ (on bottom)

$$g_{i+u}^{\alpha a_{i_1} \cdots a_{i_u}} := \begin{cases} g_i^\alpha, & \text{if } i_1 \cdots i_u \text{ is empty} \\ e_{i+u-1}(\cdots e_i(g_i^\alpha, L(i_u)) \cdots, L(i_1)), & \text{otherwise} \end{cases}$$

(recall that $i_u < \cdots < i_1$). Moreover, define $Shift(\perp, i_1 \cdots i_u, L) = \perp$.

$Recon(\mathcal{C}, P, L, A, V_A)$

1. Initially, all gates of $\mathcal{C}$ are unmarked and $R(i) := V_A(i)$, for each input wire i of $\mathcal{C}$;

2. If $\Gamma = (w_1, w_2, OR, w)$ is unmarked and both $R(w_1)$ and $R(w_2)$ were assigned, then:
   - mark Γ and
   - $R(w) := sup\{Shift(R(w_1), i_1 \cdots i_u, L), Shift(R(w_2), j_1 \cdots j_v, L)\}$,
   where $i_1 \cdots i_u$ and $j_1 \cdots j_v$ are the left and right FO-level sequences of Γ, respectively;

3. If $\Gamma = (w_1, w_2, AND, w)$ is unmarked and both $R(w_1)$ and $R(w_2)$ were assigned, then:
   - mark Γ and
   - $R(w) := Shift(R(w_1), i_1 \cdots i_u, L) \cdot Shift(R(w_2), j_1 \cdots j_v, L)$,
   where $i_1 \cdots i_u$ and $j_1 \cdots j_v$ are the left and right FO-level sequences of Γ, respectively;

4. If $\Gamma = (w, FO, w_1, \ldots, w_j)$ is unmarked and $R(w)$ was assigned, then:
   - mark Γ and
   - $R(w_i :) = e_u(R(w), P(w_i))$ for all $1 \leq i \leq j$,
   where $R(w)$ is of the form g_u^α for some u and α;

5. repeat the last three steps above until all gates get marked.

$Recon$ is illustrated in Fig. 3. We are now in a position to define our new scheme.

$KP - ABE\_Scheme\_2$

$Setup(\lambda, n, r)$: chooses a prime p, $r + 2$ multiplicative groups $G_1, \ldots, G_{r+2}$ of prime order p, a generator $g_1 \in G_1$, and a sequence of bilinear maps $(e_i : G_i \times G_1 \to G_{i+1} | 1 \leq i \leq r+1)$. Denote $g_{i+1} = e_i(g_i, g_1)$, for all $1 \leq i \leq r+1$. Then, it defines the set of attributes $\mathcal{U} = \{1, \ldots, n\}$, chooses $y \leftarrow \mathbb{Z}_p$ and $t_i \leftarrow \mathbb{Z}_p$, for each $i \in \mathcal{U}$. Finally, it outputs the public parameters $PP = (n, r, p, G_1, \ldots, G_{r+2}, g_1, e_1, \ldots, e_{r+1}, Y = g_{r+2}^y, (T_i = g_1^{t_i} | i \in \mathcal{U}))$ and the master key $MSK = (y, t_1, \ldots, t_n)$;

$Encrypt(m, A, PP)$: encrypts a message $m \in G_{r+2}$ under a non-empty set $A \subseteq \mathcal{U}$ of attributes by $E = (A, E' = mY^s, (E_i = T_i^s = g_1^{t_i s} | i \in A))$, where $s \leftarrow \mathbb{Z}_p$;

$KeyGen(\mathcal{C}, MSK)$: generates a decryption key $D = ((D(i) | i \in \mathcal{U}), P, L)$, where $(S, P, L) \leftarrow Share(y, \mathcal{C})$ and $D(i) = g_1^{S(i)/t_i}$ for all $i \in \mathcal{U}$;

$Decrypt(E, D)$: computes $m := E'/R(o)$, where $R := Recon(\mathcal{C}, P, L, A, V_A)$ and, for all $i \in \mathcal{U}$,

$$V_A(i) = \begin{cases} e_1(E_i, D(i)) = e_1(g_1^{t_i s}, g_1^{S(i)/t_i}) = g_2^{S(i)s}, & \text{if } i \in A \\ \perp, & \text{otherwise} \end{cases}$$

It is straightforward to prove the correctness of our KP-ABE_Scheme_2. As with respect to security, we have the following important result (its proof can be found in [4]).

Theorem 2. *The KP-ABE_Scheme_2 is secure in the selective model under the decisional multi-linear Diffie-Hellman assumption.*

4.2   Implementation, Complexity, and Comparisons

Translation to Graded Encoding Systems. Our KP-ABE_Scheme_2 can be translated into the graded encoding system formalism [5] exactly as in [6] and, therefore, the details are omitted. We only emphasize that:

1. the level zero in our notation corresponds to the encoding level two in [5];
2. starting with the first level (in our construction), any FO-level (and only them) corresponds to an encoding level [5], in increasing order;
3. the keys in (S, P, L) in our construction are level one encodings in [5].

Improvements. A number of improvements and extensions can be added to KP-ABE_Scheme_2. For instance, the scheme can be extended to Boolean circuits with logic gates of fan-in more than two, without increasing the size of the decryption key or of the chained multi-linear map. Such an extension could be useful in order to reduce the number of gates and the depth of a given Boolean circuit, resulting in a possible smaller decryption key.

Our KP-ABE_Scheme_2 is defined for a fixed number r of FO-levels. However, we can easily extend it to correspond to an arbitrary but upper bounded number of such levels. The main idea is to add FO-level-keys for the "missing FO-levels".

An important improvement of our scheme consists of using the FO-level-key of a FO-level as a FO-key for the first output wire of each FO-gate on that level. The main benefit of this new KP-ABE scheme consists of the fact that the number of decryption key components is decreased by the number of FO-gates.

Complexity and comparisons. The efficiency of our scheme (the improved version), in comparison with the scheme in [6] which falls in the same class of schemes as ours, is presented in the following table.

5   Attribute Multiplication-Based (AM) KP-ABE Scheme

In [9] the authors remarked that the idea in [3] can also be applied if the sub-circuits with roots consisting of gates with fan-out greater than one are multiplied. The multiplication number is exactly the number of output wires. By transforming the Boolean circuit in this way, each attribute may get multiple copies (even an exponential number of copies). The secret sharing and reconstruction is similar to the one in KP-ABE_Scheme_1. The number of decryption key components satisfies the following inequalities

$$n + r(j - 1) \leq \text{no.keys} \leq n + j^r$$

Monotone Boolean circuits with – n_1 (n_2) input gates of fan-out 1 (> 1) – q_1 (q_2) logic gates of fan-out 1 (> 1) – r FO-levels and depth ℓ	No of keys	Multi-linear map (type, size, and mult. depth)
KP-ABE scheme in [6]	$2(n_1 + n_2) +$ $3(q_1 + q_2) \leq$ $no.\,keys \leq$ $2(n_1 + n_2) +$ $4(q_1 + q_2)$	• leveled • $\dfrac{\ell(\ell + 1)}{2}$ • $\ell + 1$
KP-ABE_Scheme_2	$n_2 + q_1 + q_2 +$ $3 \leq no.\,keys \leq$ $n_2 + q_1 +$ $2q_2 + 2$	• chained • $r + 1 < \ell$ • $r + 1$

in comparison with KP-ABE_Scheme_1 where the number of keys satisfies

$$nj + n + r(j - 1) \leq \text{no.keys} \leq nj + n + j^r$$

(n is the number of input gates, r is the number of gates of fan-out greater than one, and j is the maximum fan-out of gates).

As one can see, the scheme in [9] is slightly more efficient than KP-ABE_Scheme_1 with respect to the decryption key size. However, it needs to add multiple copies of the attributes. Summing up, the scheme still remains inefficient for the entire class of Boolean circuits.

6 Conclusions

We have surveyed in this paper five KP-ABE schemes for Boolean circuits. The first scheme [8] is based on secret sharing and just one bilinear map, and works only for Boolean formulas. The second scheme [6] uses multi-linear maps and works for the entire class of Boolean circuits.

The third scheme [3] can be viewed as an extension of the scheme in [8]. The efficiency of this scheme depends on the number of FANOUT-gates and their positions in the circuit. Thus, for Boolean circuits representing multi-level access structures this scheme performs better than the one in [6]. For more "complex" Boolean circuits, the KP-ABE scheme in [6] may have a better complexity with respect to the number of decryption keys. However, we may think that the scheme in [3] acts as a bridge between the approach in [8] based on secret sharing and just one bilinear map (but limited to Boolean formulas), and the one in [6] based only on multi-linear maps (which works for general Boolean circuits).

The fourth KP-ABE scheme [4] is based on secret sharing and a particular and a special form of leveled multi-linear maps, called *chained multi-linear maps*. This scheme is more efficient than the one in [6], both in terms of the decryption key size and of the multi-linear map size.

The last scheme [9] is a slight improvement of the scheme in [3].

Currently, leveled multi-linear maps are not safe and the complexity of KP-ABE schemes from bilinear maps is still too high for the entire class of Boolean circuits. How to design secret sharing procedures for Boolean circuit still needs further research.

References

1. Bellare, M., Hoang, V.T., Rogaway, P.: Foundations of garbled circuits. In: Proceedings of the 2012 ACM Conference on Computer and Communications Security, CCS 2012, pp. 784–796. ACM, New York (2012)
2. Bethencourt, J., Sahai, A., Waters, B.: Ciphertext-policy attribute-based encryption. In: IEEE Symposium on Security and Privacy, S&P 2007, pp. 321–334. IEEE Computer Society (2007)
3. Țiplea, F.L., Drăgan, C.C.: Key-policy attribute-based encryption for boolean circuits from bilinear maps. In: Ors, B., Preneel, B. (eds.) BalkanCryptSec 2014. LNCS, vol. 9024, pp. 175–193. Springer, Cham (2015). doi:10.1007/978-3-319-21356-9_12
4. Drăgan, C.C., Țiplea, F.L.: Key-policy attribute-based encryption for general boolean circuits from secret sharing and multi-linear maps. In: Pasalic, E., Knudsen, L.R. (eds.) BalkanCryptSec 2015. LNCS, vol. 9540, pp. 112–133. Springer, Cham (2016). doi:10.1007/978-3-319-29172-7_8
5. Garg, S., Gentry, C., Halevi, S.: Candidate multilinear maps from ideal lattices. In: Johansson, T., Nguyen, P.Q. (eds.) EUROCRYPT 2013. LNCS, vol. 7881, pp. 1–17. Springer, Heidelberg (2013). doi:10.1007/978-3-642-38348-9_1
6. Garg, S., Gentry, C., Halevi, S., Sahai, A., Waters, B.: Attribute-based encryption for circuits from multilinear maps. In: Canetti, R., Garay, J.A. (eds.) CRYPTO 2013. LNCS, vol. 8043, pp. 479–499. Springer, Heidelberg (2013). doi:10.1007/978-3-642-40084-1_27
7. Gorbunov, S., Vaikuntanathan, V., Wee, H.: Attribute-based encryption for circuits. In: Boneh, D., Roughgarden, T., Feigenbaum, J. (eds.) STOC, pp. 545–554. ACM (2013)
8. Goyal, V., Pandey, O., Sahai, A., Waters, B.: Attribute-based encryption for fine-grained access control of encrypted data. In: ACM Conference on Computer and Communications Security, pp. 89–98. ACM (2006)
9. Peng, H., Gao, H.: A key-policy attribute-based encryption scheme for general circuit from bilinear maps. Int. J. Network Secur. 19(5), 704–710 (2017)
10. Ostrovsky, R., Sahai, A., Waters, B.: Attribute-based encryption with non-monotonic access structures. In: ACM Conference on Computer and Communications Security, pp. 195–203. ACM (2007)
11. Simmons, G.J.: How to (Really) share a secret. In: Goldwasser, S. (ed.) CRYPTO 1988. LNCS, vol. 403, pp. 390–448. Springer, New York (1990). doi:10.1007/0-387-34799-2_30
12. Stinson, D.R.: Cryptography: Theory and Practice, 3rd edn. Chapman and Hall/CRC, Boca Raton (2005)
13. Tassa, T.: Hierarchical threshold secret sharing. J. Cryptology 20(2), 237–264 (2007)
14. Tassa, T., Dyn, N.: Multipartite secret sharing by bivariate interpolation. J. Cryptology 22(2), 227–258 (2008)

Security of Pseudo-Random Number Generators with Input

(Invited Talk)

Damien Vergnaud[✉]

Département d'informatique de l'ENS, École normale supérieure,
CNRS, PSL Research University, 75005 Paris, France
damien.vergnaud@ens.fr

Abstract. A pseudo-random number generator is a deterministic algorithm that produces numbers whose distribution is indistinguishable from uniform. A formal security model for pseudo-random number generator with input was proposed in 2005 by Barak and Halevi. This model involves an internal state that is refreshed with a (potentially biased) external random source, and a cryptographic function that outputs random numbers from the internal state. We briefly discuss the Barak-Halevi model and its extension proposed in 2013 by Dodis, Pointcheval, Ruhault, Wichs and Vergnaud to include a new security property capturing how a pseudo-random number generator should accumulate the entropy of the input data into the internal state. This property states that a pseudo-random number generator with input should be able to eventually recover from compromise even if the entropy is injected into the system at a very slow pace, and expresses the real-life expected behavior of existing designs. We also outline some variants of this model that were proposed recently.

1 Introduction

Randomness is a key ingredient for cryptography. Random bits are necessary not only for generating keys, but are also often used in cryptographic algorithms (e.g. in probabilistic encryption schemes to achieve semantic security, in signature schemes such as DSA or ECDSA as ephemeral keys, in challenge-response protocols, as initialization vectors for block ciphers modes of operation, in side-channel countermeasures, ...).

A convenient abstraction in the design and analysis of probabilistic cryptographic algorithms is that they are given a stream of completely unbiased and independent random bits (in particular, they are unpredictable). Unfortunately, the real-world implementations of theoretically secure cryptosystems involve many failures, especially due to the fact that obtaining "perfect" randomness is a challenging task. Indeed, the physical sources that produce randomness often have non-uniform or even unknown distributions. Moreover, even if some

© Springer International Publishing AG 2017
P. Farshim and E. Simion (Eds.): SecITC 2017, LNCS 10543, pp. 43–51, 2017.
https://doi.org/10.1007/978-3-319-69284-5_4

processing is done on the randomness in order to give it a better quality, there exist many scenarios in which an attacker could be able to recover information (physical attacks, reinitialization of the system, . . .).

For these reasons, random bits in protocols are often generated by a *Pseudo-Random Number Generator* (PRNG). When this is done, the security of the scheme depends of course in a crucial way on the quality of the pseudo-randomness generated. If a user has access to a truly random bit-string, he can use a *deterministic* (or *cryptographic*) PRNG to expand this short *seed* into a longer sequence which distribution is indistinguishable from the uniform distribution to a computationally-bounded adversary (which does not know the seed) [BM82,BM84,Yao82]. The existence of PRNGs is a major open problem and in a seminal paper, Håstad, Impagliazzo, Levin, and Luby [HILL99] showed that pseudo-random generators exist if and only if one-way functions exist. In many situations, it is unrealistic to assume that users have access to secret and perfect randomness. In a PRNG with input, one only assumes that users can securely store a secret internal state and have access to some (possibly biased) random source. These PRNGs with input use a cryptographic function (*e.g.* a deterministic PRNG) to output pseudo-random numbers from the internal state and have their internal state regularly refreshed with entropy, though from this possibly biased source of randomness.

The lack of insurance about the generated random numbers can cause serious damages, and vulnerabilities can be exploited by attackers. As evidence that the use of randomness in cryptography is error-prone, one striking example is the randomness failure in the early versions of Netscape's Secure Socket Layer (SSL) encryption protocol in 1996. The protocol used pseudo-random quantities derived from a PRNG seeded with three variable values: the time of day, the process ID, and the parent process ID. These quantities have little entropy and are often relatively predictable and that version of SSL was found to be insecure [GW96]. In 2008, a similar example was revealed in Debian Linux distribution [CVE08], where a bug in the OpenSSL package[1] led to insufficient entropy gathering and to practical attacks on the SSH and SSL protocols (the only remaining source of entropy coming from the process PID, i.e. 16 bits or less of effective entropy).

There exists several digital signature schemes based on the discrete logarithm problem (e.g. DSA, ECDSA, Schnorr signature scheme, . . .). These schemes use a nonce k (or ephemeral key) for each signed message and compute g to the power k for some group element g in the underlying group. In order to illustrate the vulnerabilities of a cryptosystem to weaknesses of the underlying pseudo-random generators, Bellare, Goldwasser and Micciancio [BGM97] studied in 1997 the impact of using a linear congruential generator to generate the nonces k in the DSA signature algorithm. They showed that in this setting the scheme becomes completely insecure even if the attacker does not have access to the actual

---

[1] OpenSSL 0.9.8c-1 up to versions before 0.9.8g-9.

outputs of the generator[2]. Two famous attacks were designed later against these signature schemes that exploit only partial information on the nonces of several signatures: Bleichenbacher's attack [Ble00,MHMP13] and Howgrave-Graham and Smart's attack [HS01,NS02,NS03]. A real-world attack against these signature schemes was the recovery in December 2010 of the ECDSA private key used by Sony to sign software for the PlayStation 3 game console. This attack is due to a serious bug in the process of generating randomness for the nonces k. Another devastating flaw was revealed in August 2013, when one realized that the Java class SecureRandom could generate collisions in the nonce values used for ECDSA in implementations of Bitcoin on Android [KHL13] (if a user signs two different messages using the same ephemeral key, then the long-term ECDSA private key is immediately computable from the public key and the signatures).

In 2012, Lenstra, Hughes, Augier, Bos, Kleinjung and Wachter [LHA+12] performed a sanity check of public keys collected on the web and showed that a significant percentage of RSA public keys share a common prime factor. In an independent work, Heninger, Durumeric, Wustrow and Halderman [HDWH12] analyzed in 2012 the data from large TLS and SSH scans and identified several patterns of vulnerability: they notably found (1) repeated RSA keys due apparently to low entropy during key generation; (2) different RSA public keys sharing a common prime factor (and therefore factorable) as in [LHA+12] and (3) DSA signature keys that were used to sign two messages with the same ephemeral key (and are therefore immediately computable as above). Heninger *et al.* also presented an interesting analysis that explains the generation of these low entropy keys at boot time.

These examples, among many others, show how dramatic the consequences of using randomness of poor quality could be. They illustrate the need for precise evaluation of PRNGs based on clear security requirements.

2   Security Models

Descriptions of PRNGs with input are given in various standards [Kil11,ISO11, ESC05]. These standards identify the following core components of a PRNG with input:

- an *entropy source* which is the source of randomness used by the generator;
- an *internal state* which consists of all the parameters and variables that the PRNG uses for its operations.

---

[2] In a similar setting, Koshiba [Kos02] proved that the linear congruential generator can be used to generate randomness in the ElGamal encryption scheme (based on some plausible assumption). Fouque, Tibouchi, and Zapalowicz [FTZ13] analyzed the security of public-key schemes when the secret keys are constructed by concatenating the outputs of a linear congruential generator. Benhamouda, Chevalier, Thillard and Vergnaud [BCTV16] proposed attacks when the RSA prime factors are constructed in this way and against the RSA encryption padding described in PKCS #1 v.1.5 when a linear congruential generator is used to generate random values.

Three security notions are usually considered for this primitive:

- the *resilience* models the inability of the adversary to predict future PRNG outputs even when manipulating the entropy source;
- the *forward security* ensures that the adversary cannot get information on past outputs of the PRNG even if it is able to compromise its internal state
- the *backward security* ensures that the adversary cannot predict future outputs of the PRNG even if it is able to compromise its internal state (provided enough fresh entropy is injected into the system after the compromise).

In spite of being widely deployed in practice, PRNGs with input were not studied intensively until recently. Gutmann [Gut98], and Kelsey, Schneier, Wagner and Hall [KSWH98] gave useful guidelines for the design of secure PRNGs with input. Desai, Hevia and Yin proposed a first formal model in 2002 [DHY02] and Barak, Shaltiel and Tromer [BST03] proposed another security model where an adversary can have some control on the randomness source.

The seminal security model was proposed by Barak and Halevi in 2005 [BH05]. They proposed a new security notion, called *robustness*, which encompasses the three mentioned security notions (resilience, forward security and backward security). Under the robustness security notion, an adversary can observe the inputs and outputs of a PRNG, manipulate its entropy source, and compromise its internal state.

In 2013, Dodis, Pointcheval, Ruhault, Vergnaud and Wichs [DPR+13] extended the work of Barak and Halevi to integrate the process of accumulation of entropy into the internal state. For this purpose, they refined the notion of robustness and proposed a practical scheme satisfying it.

In Dodis *et al.* security model, a PRNG with input is a defined as triple of deterministic algorithms $\mathcal{G} = (\mathsf{setup}, \mathsf{refresh}, \mathsf{next})$. The setup algorithm generates a value denoted seed (possibly empty) which models public parameters and possibly a common reference string. The algorithm refresh given seed, an internal state and an entropy input, outputs a new internal state. The algorithm next given seed and an internal state, outputs a new internal state and some (pseudorandom) output. In the robustness security notion, Dodis *et al.* considered two adversarial entities:

- the *adversary* $\mathcal{A}$ whose task is to distinguish the outputs of the PRNG from true randomness (in a classical real-or-random security game),
- the *distribution sampler* $\mathcal{D}$ whose task is to produce inputs which have high entropy *collectively*, but somehow help $\mathcal{A}$ in breaking the security of the PRNG.

The adversary $\mathcal{A}$ always knows seed but the distribution sampler $\mathcal{D}$ does not. Actually, the robustness security property cannot be achieved if the distribution sampler knows seed for the same reason that no deterministic randomness extractor is capable to simultaneously extract good randomness from all efficiently samplable high-entropy distributions. The distribution sampler $\mathcal{D}$ is a *stateful and probabilistic* algorithm which models potentially adversarial environment where the PRNG with input is forced to operate.

The robustness property is parameterized by some parameter γ^* which is part of the claimed PRNG security. It intuitively measures the minimal "fresh" entropy in the system when security should be expected. A PRNG with input should be able to recover from internal state compromise as long as the *total* amount of fresh entropy accumulated from the distribution sampler $\mathcal{D}$ over some *potentially long* period of time crosses the threshold γ^*. Dodis *et al.* defined two properties of a PRNG with input which are simpler to analyze than the full robustness security:

- *recovering security* which (intuitively) says that if internal state is compromised (i.e. known by the adversary) but is refreshed with many input data with sufficient collective entropy (more than the threshold γ^*) resulting in some updated final state, then the output generated by next on input this state is indistinguishable from uniform;
- *preserving security* which (intuitively) says that if the internal state starts uniformly random and uncompromised, and is then refreshed with arbitrary (adversarial) input data, resulting in some final state, then the output generated by next on input this state is indistinguishable from uniform.

Dodis *et al.* showed that these two properties, taken together, imply robustness.

Dodis *et al.* showed that the elegant construction proposed by Barak and Halevi in [BH05] is not entropy accumulating (as described above) but they proposed an efficient construction which achieves this new security property. Dodis *et al.* also analyzed the two Linux PRNGs with input /dev/random and /dev/urandom [LRSV12] and showed that both are not robust. The theoretical attacks exploit notably weaknesses in the entropy estimation performed by the two PRNGs with input.

The robustness notion was notably used in 2015 to analyze the security of the Intel Secure Key hardware RNG (ISK-RNG) in [ST15] by Shrimpton and Terashima. It was also considered to study a PRNG with input based on the sponge paradigm in [GT16] by Gazi and Tessaro[3]. The notion was also considered in the setting of *backdoored* PRNG in [DPSW16].

In [DSSW14], Dodis, Shamir, Stephens-Davidowitz and Wichs extended the model from [DPR+13] to address some additional desirable security properties of PRNGs with input. The main such property is resilience to the "premature next attack". This general attack, first explicitly mentioned by Kelsey *et al.* [KSWH98] is applicable in situations in which the PRNG internal state has accumulated an insufficient amount of entropy $\gamma < \gamma^*$ and then must produce some outputs via next. In this case, the output is not pseudo-random, but now the attacker can potentially use it to recover the current internal state by brute force ("emptying" the γ bits of entropy accumulated so far). Two practical PRNGs with input were designed to overcome this premature next problem: *Yarrow* (designed by Kelsey, Schneier and Ferguson [KSF99] and used by

---

[3] Gazi and Tessaro proposed a variant of a construction proposed by Bertoni, Daemen, Peeters and Van Assche in [BDPV10] and proved that it achieves robustness in a variant of the security framework of Dodis *et al.* in the ideal permutation model.

MacOS/iOS/FreeBSD), and *Fortuna* (subsequently designed by Ferguson and Schneier [FS03] and used by Windows). These generators partition the incoming entropy into multiple entropy "pools" and use these pools at different rates when producing outputs. This strategy ensures that at least one pool will eventually accumulate enough entropy to guarantee security before it is "prematurely emptied" by a next call. Dodis *et al.* proposed a construction with provable security inspired by these designs in [DSSW14].

In [CR14], Cornejo and Ruhault studied several PRNGs with input from different popular providers, (including OpenSSL, OpenJDK, Android, IBM and Bouncy Castle) with a particular emphasis on their internal states. They formalized a framework based on the robustness security notion from [DPR+13] to capture the notion of how much of an internal state must be corrupted in order to break a PRNG with input. Using this framework, they determined the number of bits of the internal state that an attacker needs to corrupt in order to produce a predictable output for some widely used PRNGs with input. In [ABP+15], Abdalla, Belaïd, Pointcheval, Ruhault and Vergnaud extended the robustness security notion to deal with partial leakage of sensitive information. Their goal was to consider the reality of embedded devices which may be subject to side-channel attacks where an attacker can exploit the physical leakage of a device by several means (such as power consumption, execution time or electromagnetic radiation). Their new security notion, termed leakage-resilient robust PRNG with input, allows the adversary to continuously get some leakage on the manipulated data. They analyzed Dodis *et al.* construction from [DPR+13] with respect to their new stronger security model, and proved that when used with a stronger classical deterministic PRNG, it also resists leakage.

All these security models are described in details in the recent comprehensive survey by Ruhault [Ruh17] (see also [Ruh15]).

Acknowledgments. The author would like to thank his co-authors on this active and interesting research area: Michel Abdalla, Sonia Belaïd, Yevgeniy Dodis, David Pointcheval, Sylvain Ruhault and Daniel Wichs.

References

[ABP+15] Abdalla, M., Belaïd, S., Pointcheval, D., Ruhault, S., Vergnaud, D.: Robust pseudo-random number generators with input secure against side-channel attacks. In: Malkin, T., Kolesnikov, V., Lewko, A.B., Polychronakis, M. (eds.) ACNS 2015. LNCS, vol. 9092, pp. 635–654. Springer, Cham (2015). doi:10.1007/978-3-319-28166-7_31

[BCTV16] Benhamouda, F., Chevalier, C., Thillard, A., Vergnaud, D.: Easing coppersmith methods using analytic combinatorics: applications to public-key cryptography with weak pseudorandomness. In: Cheng, C.-M., Chung, K.-M., Persiano, G., Yang, B.-Y. (eds.) PKC 2016. LNCS, vol. 9615, pp. 36–66. Springer, Heidelberg (2016). doi:10.1007/978-3-662-49387-8_3

[BDPV10] Bertoni, G., Daemen, J., Peeters, M., Van Assche, G.: Sponge-based pseudo-random number generators. In: Mangard, S., Standaert, F.-X. (eds.) CHES 2010. LNCS, vol. 6225, pp. 33–47. Springer, Heidelberg (2010). doi:10.1007/978-3-642-15031-9_3

[BGM97] Bellare, M., Goldwasser, S., Micciancio, D.: "Pseudo-random" number generation within cryptographic algorithms: the DDS case. In: Kaliski, B.S. (ed.) CRYPTO 1997. LNCS, vol. 1294, pp. 277–291. Springer, Heidelberg (1997). doi:10.1007/BFb0052242

[BH05] Barak, B., Halevi, S.: A model and architecture for pseudo-random generation with applications to /dev/random. In: Atluri, V., Meadows, C., Juels, A. (eds.) ACM CCS 05: 12th Conference on Computer and Communications Security, Alexandria, Virginia, USA, pp. 203–212, 7–11 November 2005. ACM Press (2005)

[Ble00] Bleichenbacher, D.: On the generation of one-time keys in DL signature schemes. In: Presentation at IEEE P1363 Working Group meeting, November 2000

[BM82] Blum, M., Micali, S.: How to generate cryptographically strong sequences of pseudo random bits. In: 23rd Annual Symposium on Foundations of Computer Science, Chicago, Illinois, 3–5 November 1982, pp. 112–117. IEEE Computer Society Press (1982)

[BM84] Blum, M., Micali, S.: How to generate cryptographically strong sequences of pseudo-random bits. SIAM J. Comput. 13, 850–864 (1984)

[BST03] Barak, B., Shaltiel, R., Tromer, E.: True random number generators secure in a changing environment. In: Walter, C.D., Koç, Ç.K., Paar, C. (eds.) CHES 2003. LNCS, vol. 2779, pp. 166–180. Springer, Heidelberg (2003). doi:10.1007/978-3-540-45238-6_14

[CR14] Cornejo, M., Ruhault, S.: Characterization of real-life PRNGs under partial state corruption. In: Ahn, G.-J., Yung, M., Li, N. (eds.) ACM CCS 14: 21st Conference on Computer and Communications Security, Scottsdale, AZ, USA, 3–7 November 2014, pp. 1004–1015. ACM Press (2014)

[CVE08] CVE-2008-0166. Common Vulnerabilities and Exposures (2008)

[DHY02] Desai, A., Hevia, A., Yin, Y.L.: A practice-oriented treatment of pseudo-random number generators. In: Knudsen, L.R. (ed.) EUROCRYPT 2002. LNCS, vol. 2332, pp. 368–383. Springer, Heidelberg (2002). doi:10.1007/3-540-46035-7_24

[DPR+13] Dodis, Y., Pointcheval, D., Ruhault, S., Vergnaud, D., Wichs, D.: Security analysis of pseudo-random number generators with input: /dev/random is not robust. In: Sadeghi, D., Gligor, V.D., Yung, M. (eds.) ACM CCS 13: 20th Conference on Computer and Communications Security, Berlin, Germany, 4–8 November 2013, pp. 647–658. ACM Press (2013)

[DPSW16] Degabriele, J.P., Paterson, K.G., Schuldt, J.C.N., Woodage, J.: Backdoors in pseudorandom number generators: possibility and impossibility results. In: Robshaw, M., Katz, J. (eds.) CRYPTO 2016. LNCS, vol. 9814, pp. 403–432. Springer, Heidelberg (2016). doi:10.1007/978-3-662-53018-4_15

[DSSW14] Dodis, Y., Shamir, A., Stephens-Davidowitz, N., Wichs, D.: How to eat your entropy and have it too – optimal recovery strategies for compromised RNGs. In: Garay, J.A., Gennaro, R. (eds.) CRYPTO 2014. LNCS, vol. 8617, pp. 37–54. Springer, Heidelberg (2014). doi:10.1007/978-3-662-44381-1_3

[ESC05] Eastlake, D., Scoreder, J., Crocker, S.: RFC 4086 - Randomness Requirements for Security, June 2005

[FS03] Ferguson, N., Schneier, B.: Practical Cryptography. Wiley, New York (2003)

[FTZ13] Fouque, P.-A., Tibouchi, M., Zapalowicz, J.-C.: Recovering private keys generated with weak PRNGs. In: Stam, M. (ed.) IMACC 2013. LNCS, vol. 8308, pp. 158–172. Springer, Heidelberg (2013). doi:10.1007/978-3-642-45239-0_10

[GT16] Gaži, P., Tessaro, S.: Provably robust sponge-based PRNGs and KDFs. In: Fischlin, M., Coron, J.-S. (eds.) EUROCRYPT 2016. LNCS, vol. 9665, pp. 87–116. Springer, Heidelberg (2016). doi:10.1007/978-3-662-49890-3_4

[Gut98] Gutmann, P.: Software generation of practically strong random numbers. In: Proceedings of the 7th USENIX Security Symposium (1998). http://www.cypherpunks.to/peter/06_random.pdf

[GW96] Goldberg, I., Wagner, D.: Randomness and the netscape browser. Dr. Dobb's J. **2**(1), 66–70 (1996)

[HDWH12] Heninger, N., Durumeric, Z., Wustrow, E., Halderman, J.A.: Mining your PS and QS: detection of widespread weak keys in network devices. In: Kohno, T. (ed.) Proceedings of the 21th USENIX Security Symposium, Bellevue, WA, USA, 8–10 August 2012, pp. 205–220. USENIX Association (2012). https://www.usenix.org/conference/usenixsecurity12/technical-sessions/presentation/heninger

[HILL99] Hastad, J., Impagliazzo, R., Levin, L.A., Luby, M.: A pseudorandom generator from any one-way function. SIAM J. Comput. **28**, 12–24 (1999)

[HS01] Howgrave-Graham, N., Smart, N.P.: Lattice attacks on digital signature schemes. Des. Codes Cryptogr. **23**(3), 283–290 (2001)

[ISO11] Information technology - Security techniques - Random bit generation. ISO/IEC18031:2011 (2011)

[KHL13] Kim, S.H., Han, D., Lee, D.H.: Predictability of android OpenSSL's pseudo random number generator. In: Sadeghi, A.-R., Gligor, V.D., Yung, M. (eds.) ACM CCS 13: 20th Conference on Computer and Communications Security, Berlin, Germany, 4–8 November 2013, pp. 659–668. ACM Press (2013)

[Kil11] Killmann, W., Schindler, W.: A proposal for: Functionality classes for random number generators. AIS 20/AIS31 (2011)

[Kos02] Koshiba, T.: On sufficient randomness for secure public-key cryptosystems. In: Naccache, D., Paillier, P. (eds.) PKC 2002. LNCS, vol. 2274, pp. 34–47. Springer, Heidelberg (2002). doi:10.1007/3-540-45664-3_3

[KSF99] Kelsey, J., Schneier, B., Ferguson, N.: Yarrow-160: notes on the design and analysis of the yarrow cryptographic pseudorandom number generator. In: Heys, H., Adams, C. (eds.) SAC 1999. LNCS, vol. 1758, pp. 13–33. Springer, Heidelberg (2000). doi:10.1007/3-540-46513-8_2

[KSWH98] Kelsey, J., Schneier, B., Wagner, D., Hall, C.: Cryptanalytic attacks on pseudorandom number generators. In: Vaudenay, S. (ed.) FSE 1998. LNCS, vol. 1372, pp. 168–188. Springer, Heidelberg (1998). doi:10.1007/3-540-69710-1_12

[LHA+12] Lenstra, A.K., Hughes, J.P., Augier, M., Bos, J.W., Kleinjung, T., Wachter, C.: Public keys. In: Safavi-Naini, R., Canetti, R. (eds.) CRYPTO 2012. LNCS, vol. 7417, pp. 626–642. Springer, Heidelberg (2012). doi:10.1007/978-3-642-32009-5_37

[LRSV12] Lacharme, P., Röck, A., Strubel, V., Videau, M.: The linux pseudorandom number generator revisited. Cryptology ePrint Archive, Report 2012/251 (2012). http://eprint.iacr.org/2012/251

[MHMP13] De Mulder, E., Hutter, M., Marson, M.E., Pearson, P.: Using Bleichen-bacher's solution to the hidden number problem to attack nonce leaks in 384-Bit ECDSA. In: Bertoni, G., Coron, J.-S. (eds.) CHES 2013. LNCS, vol. 8086, pp. 435–452. Springer, Heidelberg (2013). doi:10.1007/978-3-642-40349-1_25

[NS02] Nguyen, P.Q., Shparlinski, I.: The insecurity of the digital signature algorithm with partially known nonces. J. Cryptol. **15**(3), 151–176 (2002)

[NS03] Nguyen, P.Q., Shparlinski, I.E.: The insecurity of the elliptic curve digital signature algorithm with partially known nonces. Des. Codes Cryptogr. **30**(2), 201–217 (2003)

[Ruh15] Ruhault, S.: Security analysis for pseudo-random numbers generators. (Analyse de Sécurité des Générateurs Pseudo-Aléatoires). Ph.D. thesis, École Normale Supérieure, Paris, France (2015). https://tel.archives-ouvertes.fr/tel-01236602

[Ruh17] Ruhault, S.: Sok: security models for pseudo-random number generators. IACR Trans. Symmetric Cryptol. **2017**(1), 506–544 (2017)

[ST15] Shrimpton, T., Terashima, R.S.: A provable-security analysis of intel's secure key RNG. In: Oswald, E., Fischlin, M. (eds.) EUROCRYPT 2015. LNCS, vol. 9056, pp. 77–100. Springer, Heidelberg (2015). doi:10.1007/978-3-662-46800-5_4

[Yao82] Yao, A.C.-C.: Theory and applications of trapdoor functions (extended abstract). In: 23rd Annual Symposium on Foundations of Computer Science, Chicago, Illinois, 3–5 November 1982, pp. 80–91. IEEE Computer Society Press (1982)

Securing the Foundations of Democracy

Peter Y.A. Ryan[✉]

University of Luxembourg, Luxembourg City, Luxembourg
peter.ryan@uni.lu

Abstract. Recent events have highlighted numerous threats to democracy, in particular the 2016 US presidential election is mired in controversy. Allegations of Russian interference with the campaigns, in particular hacking and selective leaking of emails from the Democratic campaign management, possible hacking of electronic voting and tabulating. Alongside this we have challenges to democratic debate due to "fake news", information bubbles, the chilling effect of mass surveillance etc. All of this suggests that we need to have a major rethink of how democracy should function effectively in the digital age.

In a short article we cannot hope to address all of these threats, but rather we focus on just one aspect, arguably the keystone of democracy: making secure the conduct of elections. In particular we outline approaches to making elections verifiable and accountable, while guaranteeing ballot privacy and coercion resistance.

1 Introduction

The election of Donald Trump in 2016 to the most powerful office on the planet brought into sharp relief the strains that digital technologies are placing on the democratic process. In theory such technologies could enrich the democratic process by for example facilitating the dissemination of information and fostering debate. In practice we have seen that such technologies open up new and poorly understood threats. In the case of the US presidential election we have witnessed hacking of the email servers of the Democratic campaign committee, of registers of voters and apparent attempts to hack voting and tabulation machines.

Besides all these threats to the collecting, recording and counting of votes, we are seeing a raft of threats to the surrounding processes supporting the dissemination of news and information as well as informed debate. We hear of the prevalence of *fake news* and the rise of news disseminated via social media, resulting in *information bubbles* that serve to reinforce prejudices and preconceptions. All of this undermines the informed debate and decision making that should be the bedrock of a healthy democracy.

Clearly, in a short, rather technical, paper we are not going to able to address all of these issues, rather we just attempt to overview some approaches that have emerged from the information assurance community to address the security of the election process. In particular I will give a high-level overview of approaches

© Springer International Publishing AG 2017
P. Farshim and E. Simion (Eds.): SecITC 2017, LNCS 10543, pp. 52–66, 2017.
https://doi.org/10.1007/978-3-319-69284-5_5

that go under the heading of *end-to-end verifiable* (E2E V) or sometimes *fully auditable* schemes. Here to goal is to provide strong guarantees that all legitimately cast votes are accurately counted while preserving ballot privacy, receipt freeness and, ideally, coercion resistance, while keeping trust assumptions to a minimum. After giving an outline of the design principles behind E2E V schemes, I will go into more detail on a new scheme, called Selene, that seeks to make the voter verification much more transparent and intuitive, while maintaining coercion mitigation.

1.1   Trust Assumptions in Conventional Voting Systems

Conventional voting systems typically require a significant level of trust to be placed in various components, which may be technological or human. Old fashioned voting with paper ballots and hand-counting involve trust in the people and procedures handling the ballot boxes, doing the counting etc. This can be partly offset by allowing independent observers but these will not be infallible and some level of trust still needs to be placed in their competence and honesty. Nonetheless, it can usually be argued that pulling off large-scale, fraud in a way that is likely to be undetected is extremely hard.

In the case of DREs however it is clear to anyone with even a modest understanding of the fragility of software and network security that large-scale, virtually undetectable fraud is quite easy to pull off, by simply tweaking a few lines of code. Indeed at the time of writing the DefCon conference staged demonstrations of hacking of various US voting machines, and showed that in some cases hackers could take over a machine in only minutes.

The response of most security experts to the use of voting computers is either, as was the case in the Netherlands, to demand that they be banned, or, in the US, to insist that any DRE must be supplemented by a Voter Verified Paper Audit Trail (VVPAT). This is essentially a printer at the side that prints the voters choice to paper, that can be checked by the voter and confirmed if correct. As long as the resulting paper audit trail is well curated, this creates a record that can be used for example for Risk Limiting Audits, [17]. If a link is maintained between each paper ballots and its digital representation used in the electronic tally then highly efficient *comparison audits* are possible: a typically very small, random sample of the ballots can be taken and if for all ballots the paper and electronic representations agree then a high level of confidence in the declared result can be obtained.

Clearly, if we are prepared to completely trust an authority to handle the votes correctly and ensure ballot privacy then we trivially solve the problem. However, nobody should be comfortable placing such trust for such a critical service, and for cryptographers and security specialists having to place such heavy trust is an anathema. The goal then is to try to make the processing of the votes as transparent as possible while respecting the privacy of votes. Steering a course between these conflicting requirements with minimal trust assumptions is what makes this an immensely challenging topic. Add to this the requirement that the system should be extremely easy to use and understand by every member of the electorate, who might use the system once every few years, and the fact that voters may cooperate with coercers or vote-buyers and you have arguably the biggest challenge facing information security engineers.

And we haven't even got onto the challenges of internet voting yet, with all the inherent insecurity of the internet, dangers arising from corrupted client devices, and the impossibility of preventing coercers observing the voters.

What cryptographers have sought to do, in numerous proposed schemes, is to use the rich toolkit of modern cryptography to make the execution of the election as transparent as possible. Such schemes seek to ensure that any errors or corruption in the handling of votes are detectable in a way that is, as far as possible, observable and verifiable by all. The difficulty with observing a conventional voting system is that you are monitoring an ephemeral process, if you miss some sleight of hand then the evidence is gone. In other words, in the terminology of Stark and Wagner, [18], we seek to make elections *evidence-based*: as the election unfolds, the system and authorities are required to generate sufficient evidence, of sufficient quality, to convince any reasonable sceptic, above all, the losers of the correctness of the announced result.

2   End-to-End Verifiable Schemes

In this section I give a very high-level indication of the key ideas behind the E2E V approach. The goal of such schemes is to enable each voter to be able to confirm to their own satisfaction that the vote that they cast is accurately included in the final tally. Immediately the astute reader will see that this is going to be tricky: if we provide ways to convince the voter that their vote is counted how are we going to avoid this being used to convince a coercer or vote-buyer? This is indeed immensely delicate, and this is where the magic of modern crypto comes to our aid.

The key idea is that when the vote is cast, a form of receipt is generated that carries a suitably encrypted or encoded representation of the vote. The voter gets to keep a copy of this receipt, or *protected ballot* as Rivest has suggested it be called. Copies all such receipts are posted to a *Public Bulletin Board*, (PBB). The PBB needs some explanation: it should have the following properties:

- it should be visible to all, and all should guarantee a consistent view of its contents to all,
- anything posted to the WBB should be guaranteed to remain posted and unaltered, i.e. it should be append-only,
- only appropriate authorities should be able to post items.

The voter gets to retain a copy of the encrypted vote which she can later confirm is correctly posted to the Web Bulletin Board (WBB). All the posted, encrypted ballots are then anonymously tabulated, either using mixes and decryption or exploiting homomorphic properties of the encryption to tabulate under encryption and then decrypt the result. The point of encrypting the vote is of course to ensure that even of the voter shows it to someone else the privacy of the vote remains.

A number of E2E V schemes have been proposed and some even implemented and deployed, for example, prominent in-person schemes include Prêt à Voter [15] Wombat [2] and Scantegrity II [16], StarVote [8], while internet schemes include Helios https://vote.heliosvoting.org/, Civitas [5] and Pretty Good Democracy [14].

2.1 Verifiable Tabulation

Once we have an agreed set of encrypted votes, the extraction of the tally in a veritable fashion while ensuring ballot secrecy involves subtle but well-understood cryptographic techniques. For example, the set of encrypted votes can be put through a number of verifiable mixes to ensure anonymity and then verifiably decrypted. The result is the set of decrypted votes as cast, which can be counted up by anyone, but with any link back to the original receipts obliterated. An alternative approach is to exploit homomorphic properties possessed by many probabilistic encryption algorithms: exponential ElGamal and Paillier enjoy additive homomorphic properties: the product of encrypted plaintexts equals the encryption of the sum of the plaintexts:

$$\prod_i \{x_i\}_{PK} = \{\sum_i x_i\}_{PK}$$

This is very handy property for voting systems as it allows us to sum up votes under the wraps of encryption and then just decrypt the final counts in one shot. No individual encrypted ballots are revealed so ensuring ballot secrecy. To take a simple example, suppose that we are running a referendum, we encode a *yes* vote as a $+1$ and a *no* vote as a -1. To compute the result we take the product of the encrypted votes and then decrypt the result, if this is positive that the yeas have it, if negative then the nays carry the day. More complex elections, for example, with multiple candidates, can similarly be conducted with suitable encodings of the votes.

2.2 Ballot Auditing

The really interesting, and challenging part of designing an E2E V voting system is in the creation of the encryptions of the votes. It is essential that the voter be confident that his or her vote is correctly encrypted in the receipt, and this assurance must be provided in a way that cannot be conveyed to another party. It is far from obvious how a voter is to be sure that random string of symbols printed on the receipt is a correct encoding of the intended vote. Most schemes try to tackle this with some form of *cut and choose* protocol: the voting machine commits, in print say, to a number of encryptions, k say, of the vote, and the voter gets to chose to audit $k - 1$ of these. If all the audited encryptions turn out to hold the correct vote then this provides assurance that remaining one is also correct, and this can now be cast with some confidence.

An alternative approach, referred to as *Benaloh challenges*, [3], is to do something similar but in a sequential fashion: the device generates an encryption of the vote and the voter is now given a choice as to whether to audit or cast this ballot. If she audits, and is happy with the result, she obtains a fresh encryption and again has the choice: audit or cast. This can go on in principle indefinitely until she is sufficiently confident the encryption device is behaving correctly, at which point she casts the vote and the process is complete.

An obvious question is: why not just audit one encryption, and cast this if it correct? The answer to this is that auditing typically produces a proof of the plaintext, which could then be presented to a coercer or vote-buyer. A highly ingenious scheme, MarkPledge [1], does allow the cast ballot to be audited, but this involves this voter participating in an interactive zero-knowledge proof protocol with the device in the booth. The real, interactive proof transcript for the voter's choice, generated with the voter, is masked by fake proof transcripts for the other candidates. Only the voter, who participated in the creation of the receipt in the booth, knows which was the real interactive proof. The scheme is technically quite brilliant, but the resulting complexity and lack of usability prevented wide-scale uptake.

In Prêt à Voter, printed ballot forms are generated with the candidates listed in a randomised order. Each ballot has an independently generated order and carried an encrypted representation of the order. In the privacy of the booth, the voter extracts the form from a tamper proof envelope and applies the appropriate marks against the candidate(s) of choice, an X or ranking etc. The plaintext list of the candidates is detached and destroyed. The result is an receipt carrying the vote in encrypted form: without knowledge of the order in which the candidates were presented in this particular ballot the vector of marks cannot be interpreted.

When presented with a ballot form, sealed in an envelope, the voter can opt either to cast her vote using the ballot or to audit the ballot. If she adopts to audit, and is happy with the outcome, she takes another form and again has the choice, Benaloh style. Later, during tabulation a threshold set of Tellers cooperate to extract the vote, taking appropriate care to protect the privacy of the vote.

This approach has some appealing features, notably:

- the vote is not communicated to any device, so sidestepping side-channel threats.
- ballot auditing is very clean and privacy preserving: you simply audit the blank form, if it is well-formed in the sense that the plaintext printed order agrees with the encrypted order then a vote cast with this form would be correctly encrypted.

The second point means that Prêt à Voter ballot auditing provides strong dispute resolution: there is no question of whether the fault lies with the voter, the ballot for is either well-formed or it is not, and this is wholly independent of the voter and the vote. Furthermore, this means that ballot audits can be performed by anyone: we can have independent auditors and observers performing random audits on forms, in addition to the audit performed by the voters.

3   Public Acceptance of E2E V Schemes

The general approach sketched above has a lot of technical merit and offers high assurance of accuracy of the tally along with guarantees of ballot privacy and coercion resistance with minimal trust assumptions. The assurance arguments are rather subtle though, and some people object to the use of crypto in voting on the grounds that the majority of the electorate will not really understand it and its role. People are often troubled at not being able to identify their vote in the clear in the tally, and seem unconvinced when it is pointed out that if this capability were to be provided it would open the system up to coercion and vote buying.

Ironically, the fact that errors or corruption are made detectable often does not seem to inspire trust. A good scheme should be both trustworthy and trusted. All too often we see commercial schemes that are not trustworthy apparently trusted and conversely highly trustworthy schemes which fail to inspire trust. While developing and trialling Prêt à Voter colleagues at the University of Surrey conducted focus groups. The groups were given a description of the scheme and its security guarantees and were asked what they thought of it. Many answered along the lines of: "it is all very well offering a scheme that can detect when things go wrong, but surely it would be better to design one that cannot go wrong".

All of these considerations suggest that it is interesting to explore the possibility of achieving some form of verifiability without the use of crypto. An early example of this is the article of Randell and Ryan [11] that uses scratch strips as an analogue of crypto. Another fine example is Rivest's ThreeBallot system [12].

Another objection often raised against such schemes is the point that verifiable does not automatically mean verified. For an election to be deemed verified we must be able to show that voters and observers did indeed perform the checks in sufficient numbers and with sufficient diligence. It is essential therefore that the various checks be as easy to perform as possible and well motivated. Furthermore we need reliable ways to monitor the levels of checking. A question that may spring to mind at this point is: why not just automate the checks? The answer to this is that if we automate them then we are thrown back into having to place trust in the processes performing the checks. Our goal here is to make the electorate themselves the bedrock on which the trust is based.

4   Related Work

E2E verifiable voting now has quite a long and rich literature, with many schemes having been proposed, both for in-person and remote, e.g. internet voting. Here we will just mention some of the most closely related schemes.

The most notable verifiable internet voting scheme is Adida's Helios, https://vote.heliosvoting.org/. Helios is not receipt-free, but recently the Belenios RF scheme, [6], has been proposed to provide receipt freeness.

Juels *et al.* [9] proposed a formal definition of coercion resistance and a credential-based mechanism to achieve this. The Civitas system, [5], http://www.cs.cornell.edu/projects/civitas/, implements this approach, with some enhancements.

The idea of voters having a private tracking number with which they can look up their vote in the clear on a bulletin board appears to go back the Schneier's "Applied Cryptography" book in which he suggests that voters choose a password to identify their vote. Much later the idea is revived for use in voting during ANR (Agence National de la Recherche) funding committee meetings. A scheme that has some similarities to Selene in that votes appear in the clear alongside identifying number, is Trivitas, [4]. Here, however, the clear-text votes appear on the bulletin board at an intermediate step, followed by further mixing and filtering. Hence the voters do not verify their vote directly in the final tally.

5   Selene

In this section I provide an overview of a new scheme that aims to provide voter verifiability but in a much more intuitive way, and which avoids voters handling encrypted receipts. A full description can be found at [13]. The scheme is based on an old and simple idea: voters have a private tracker number which allows them to identify their vote in the tally on the PBB. Earlier we remarked that this poses obvious problems in terms of coercion and vote buying, however the Selene scheme introduces some new twists that at least mitigates these issues.

Such an approach provides voters with a very simple, direct and easy-to-understand way to confirm that their vote is present and correct in the tally, but we must ensure that voters get distinct trackers and, as remarked above, there is a danger of coercion and vote buying. The first is an issue if, for example the system could identify two voters likely to vote the same way and assign them the same tracker. In this case it just posts one vote against this tracker and is free to insert in the tally another vote of its own choice.

The second danger is that a coercer requires the voter to hand over her tracker to allow him to check how she voted. Notice though that in this style of attack the coercer must request that the tracker be handed over before the results are published. If he asks after the trackers and votes have been published the voter has the opportunity to pick an alternative tracker pointing to the coercer's vote and claim it as her own. It is this observation that we exploit to counter this threat: we arrange for the voters to learn their tracker numbers only after the information has been posted to the WBB. The Selene scheme addresses both of these shortcomings: by guaranteeing that voters get unique trackers and arranging for voters to learn their tracker only at some time after the votes and corresponding tracking numbers have been posted (in the clear).

The hope is that by putting the crypto under the bonnet, voters, election officials etc. may find such a scheme more acceptable that conventional E2E verifiable schemes. The scheme is also interesting in that it appears to shift the trust model for voter devices: in usual E2E schemes we need to worry about the voter's device encrypting the vote correctly. As observed above, this typically necessitates complicating the voting experience with Benaloh challenges, or similar ballot assurance mechanisms. Now voters get to check their vote in the clear, a misbehaving device can be detected more readily, resulting in a simpler voting ceremony. The downside of this is that, in the event that a voter contests the posted vote, it is harder to determine which is at fault, the system or the voter. This issue, usually referred to as *dispute resolution* is a further desirable property of a good voting system, and we will return to this later.

A possible problem with the basic scheme, pointed out by Bill Roscoe, is that a coerced voter might by mis-chance choose the coercer's tracking number when she is deploying her coercion evasion strategy. Perhaps even more worrying is the possibility that the coercer will simply claim, falsely, that the tracker revealed by the voter is his and hence he "knows" that voter has not revealed her true tracker. This puts the voter in a very awkward situation.

In large elections with a small number of candidates the odds of lighting on the coercer's tracker will typically be small (unless the coercer is backing a serious loser), but even the remote possibility may be worrying to some voters. However, the other scenario: the coercer claiming, falsely, that the tracker is his, places the voter in a difficult situation. Note that this can only arise of the coercer is himself a voter.

It is not immediately obvious how to counter this danger, but an enhancement to the basic scheme which counters this possibility is described in [13]; however it comes at a cost of a less transparent tally.

The Selene scheme is in any case targeted at low coercion threat environments. We argue that, in such contexts, the benefits arising from the greater degree of transparency outweigh a rather remote and mild threat. In any event, it can be shown that the scheme provides receipt-freeness.

5.1   Selene as an Add-On

It is interesting to observe that the Selene constructions could in many cases be added to an existing voting scheme, one without any verification features or perhaps one having conventional E2E verification involving encrypted receipts. Indeed, in some cases it could even be retro-fitted to an election that had already taken place. Suppose that a Helios vote had been conducted and contested. The trapdoor commitments to the trackers could be generated and associated to the voters as described above and the mixes and decryptions performed afresh. For this to work, the base scheme must use encryption such that we can run a parallel shuffle with the corresponding encrypted trackers. Indeed, in our presentation below we will abstract away from details of exactly how votes are cast, validated etc.

6   The Set-up Phase

The EA creates the threshold election key and keys share. Ideally this should be in a distributed, dealerless fashion [7]. When voters register for the election we assume that they, or more precisely their devices, create a fresh, ephemeral trapdoor key pair. For the purposes of this paper we will leave aside the question of how voters and ballots are authenticated. It might be for example that voters all have signing keys, or they are provided with some form of credential.

We now describe the construction whose goal is to:

– ensure that each voter has a unique tracker committed to them,
– and inform voters of their tracking numbers in a way that provides them with high confidence that it is correct but allowing them to deny it if coerced.

6.1 Distributed Secret Assignment of Tracker Numbers

The tracking numbers could be short strings of digits, but could also be consecutive numbers $1, 2, \ldots$. The Election Authority (EA) publicly creates a list of the tracking numbers n_i and posts this to the PBB. Everyone can confirm that the elements of the list are pairwise distinct. EA now computes g^{n_i} (to ensure that the resulting values fall in the appropriate subgroup) and the ElGamal encryptions under the Teller's threshold public key PK_T of the g^{n_i}: $\{g^{n_i}\}_{PK_T}$ and posts these terms to the WBB:

$$n_i, \ g^{n_i}, \ \{g^{n_i}\}_{\mathsf{pk}_T}$$

These initial encryptions could be trivial, i.e. using a known randomisation such as $r = 1$ to allow universal verifiability of this step. Mix Tellers now put the encrypted terms through a sequence of verifiable, re-encryption mixes to yield:

$$\{g^{n_{\pi(i)}}\}'_{\mathsf{pk}_T}$$

where π denotes the permutation that results from the sequence of permutations applied by the mix tellers, and $\{X\}'$ denotes re-encryption of $\{X\}$. These are now assigned to the voters' IDs (or perhaps pseudo-IDs):

$$(\mathsf{ID}_i, \{g^{n_{\pi(i)}}\}'_{\mathsf{pk}_T})$$

Thanks to the multiple shuffles, the assignment of these numbers to the voters is not known to any party, only a collusion of all the mix Tellers could determine the assignment. Note also that as these are verified mixes, as long as all the input numbers are unique the assigned (encrypted) numbers will be unique to each voter.

6.2 Generation of the Tracker Number Commitments

We now need to generate, for each voter, the trapdoor commitments to the tracker. [13] gives a rather elaborate, distributed construction, but here we give a simpler construction based on a suggestion from D Wikström that uses calls to general purpose, verifiable mix net, such as Verificatum, see http://www. verificatum.com/. Using the parallel mixing facility we can generate for voter V_i a pair of ElGamal ciphertexts:

$$(u_i, v_i) = (\{g^{r_i}\}_{PK_T}, \{h_i^{r_i}\}_{PK_T})$$

We now form:

$$\{g^{n_i}\}_{PK_T} \cdot \{h_i^{r_i}\}_{PK_T} = \{g^{n_i} \cdot h_i^{r_i}\}_{PK_T}$$

and verifiably decrypt this to give the trapdoor commitment $g^{n_i} \cdot h_i^{r_i}$. The g^{r_i} term is kept encrypted and secret.

6.3 Voting

Voter V_i casts her vote using a plaintext aware encryption scheme:

$$(\{\mathsf{Vote_i}\}_{\mathsf{pk_T}}, \Pi_i)$$

The plaintext awareness is needed to prevent an attacker copying, re-encrypting and casting a previously cast vote as his own. In conjunction with Selene such a copying attack would be particularly virulent: the attacker copies the victim's vote and casts it as his own, and when the votes and trackers are revealed he sees exactly how the victim voted. It is also advisable to post the votes only once voting has closed.

The eligibility and validity of ballots is checked and, if valid, the encrypted votes are posted to the PBB to give a list of tuples on the WBB:

$$(\mathsf{ID}_i, \{g^{n_{\pi(i)}}\}_{\mathsf{pk_T}}, (h_i^{r_i} \cdot g^{n_{\pi(i)}}), (\{\mathsf{Vote_i}\}_{\mathsf{pk_T}}, \Pi_i))$$

6.4 Mixing and Decryption

Once voting has finished, for each row on the WBB, the validity of the ballot is checked for eligibility and well-formedness, according to the rules of the scheme. For valid ballots the pair of encryptions of the vote and the tracker are extracted and passed to the mixing process. This gives pairs of the form:

$$(\{g^{n_{\pi(i)}}\}_{\mathsf{pk_T}}, \{\mathsf{Vote_i}\}_{\mathsf{pk_T}})$$

These are now put through a verifiable, parallel shuffles, e.g. [10] or using Verificatum. Once this is done, a threshold set of the Tellers perform a verifiable decryption of these shuffled pairs. All of these steps along with the proofs are posted to the WBB. Thus, finally we have a list of pairs: tracking number, vote:

$$(g^{n_{\pi(i)}}, \mathsf{Vote_i})$$

from which the tracker/vote pair can immediately be derived: $(n_{\pi(i)}, \mathsf{Vote_i})$.

6.5   Notification of Tracker Numbers

Once the trackers and votes have been made available on the WBB for a sufficient period for the voters to note any alternative trackers as may be required to parry any attempted coercion, the EA sends the voter V_j their share of the g^{r_i} over a private channel:

$$T_j \rightarrow V_i : \; g^{r_j}$$

The tracker commitment terms can be thought of as the second term of an ElGamal ciphertext, with the first term, the g^r term, kept hidden. On receipt of the α term the voter, or more precisely her device, can combine this with the tracker commitment term, the β term, to form the ElGamal encryption of her tracker under her trapdoor key.

$$(\alpha, \beta) := (g^r, h^r \cdot m)$$

The device can now perform the decryption, using the trapdoor key, and reveal the tracker.

Rather surprisingly, the α term is sent to the voter without any proof of origin or authenticity. The reason that we can do this is that an adversary with bounded computational power and not possessing the relevant trapdoor key, and even if colluding with all the Tellers, has only negligible probability of constructing an α term that opens up to a valid tracker different from the true tracker of the voter. Avoiding authenticating these notifications is more user-friendly because such communications have to be deniable and should be faked by the voter in case of coercion. Designated Verifier Signatures would be a way to sidestep such coercion threats, but they would significantly complicate the user steps in the event of coercion. The precise statement and proofs can be found in [13].

Note also that for the privacy of the tracking numbers we do not really need to encrypt the g^{r_i} terms as the trackers are still protected by the encryption under the voter's PK. However, it is still important to send these terms to the voter over a private channel to ensure that they are deniable. A possibility, suggested by D Wikström, is for the voter to send an encrypted blinding factor at the same time as casting the vote. This is then used to blind the α term when it is communicated to the voter.

6.6   Coercion: Threats and Mitigation

We have described how a voter can wrong-foot a coercer who demands that she reveal he tracker number, but what about a more insistent coercer who demands that she further reveal the alpha term, and that she demonstrate how this reveals the claimed tracker when input into her device. This is where the flip side of the construction comes into play: the voter, or more precisely her device, possessing the trapdoor key, can easily compute an alternative $(g^{r_i})'$ term that will decrypt to an alternative, valid tracker of her choice. Suppose that she wants her commitment to decrypt to the tracker value m^*, she inputs this to her device along with the commitment value β_i and the device, with knowledge of the trapdoor key x, computes the fake term α':

$$\alpha' = \left(\frac{\beta_i}{m^*} \right)^{x^{-1}}$$

This still leaves a coercion possibility: the coercer can demand to observe the receipt of the α. The α terms can be sent at randomized times forcing the coercer to monitor the voter's communications. However, the possibility of receiving the α term while the coercer is present, might still be discouraging for the voter.

A possibility to circumvent this is to provide a private channel to contact the voting authorities to request that the fake $(g^{r_i})'$ term that the voter has calculated be communicated back to her. This has spin-off effect of encouraging voters to notify the authorities of coercion.

6.7 Dispute Resolution

Dispute resolution, the ability for a judge to identify the cheating or malfunctioning component or party when an error is reported, is quite hard to achieve, especially in the internet voting context. Disputes could arise in performing Benaloh challenges for example: the audited ballot opens to reveal candidate A but the voter claims to have input B. There is no immediate way to distinguish between the device cheating or the voter lying, mis-typing or mis-remembering. It is essential that a well-designed system be able to make such distinctions reliably and in a fashion that can be proven. In the absence of such a property the system will be open to attacks attempting to discredit it or the election in question: e.g. many voters reporting fake complaints.

In Selene this could be tricky if a voter claims that the vote corresponding to their tracker is not what they cast. But this is a problem with the tracking number approach anyway. We could start to resolve this but encrypting the posted vote and tracker and performing a *plaintext equivalence text* against the cast encrypted vote and tracker appearing against the voter's Id before the mixing. If the tracker number do not agree this suggests that the voter is mistaken about her tracker. If the votes do not agree there has been a problem with the parallel mixing. If both agree, and the voter continues to insist that the vote is wrong, then it is possible that her device was corrupted and performed the encryption wrongly. This would all have to be performed with great care and suitable controls, and presumably *in camera* to avoid introducing coercion opportunities.

7 Selene II

We pointed out earlier that a coerced voter might have the misfortune of choosing the coercer's tracking number, or the coercer simply claims, falsely, that this is his tracker. In mild coercion threat contexts we may be able to ignore this issue, but if the threats of the coercer are sufficiently unpleasant this possibility could be enough to deflect the voter from voting her intent. The paper [13] provides a construction that provides voters with a set of alternative trackers, each pointing to one of the candidates, in such a way that these trackers are unique to her. If coerced she simply points to the tracker from this set that points to the coercers requested candidate, and now the coercer cannot claim ownership of this tracker. The tally board will now contain $c \cdot v$ additional tracking numbers, where c is the number of the candidates and v is the number of the voters. These will give one extra vote per candidate per voter which has to be subtracted in the tally. This is ok for simple plurality style elections, but not for more elaborate social choice functions, at least not without some adaptation. This aspect of the scheme is reminiscent of Rivest's *ThreeBallot* [12].

8   Conclusions

Democracy is under severe strain, and much of this arises from the introduction of digital technologies into elections, the media, social networks etc. In this paper I have focussed on just one small aspect of these threats: securing the casting and counting of votes. Modern cryptography and information security has made significant strides over the last decade or so in devising protocols and procedures to make the conduct of elections more transparent and auditable. These range from ensuring that all systems provide a well curated paper audit trail, enabling risk limiting audits, to the use of cryptographic techniques in E2E V schemes.

In particular I presented a new voting protocol, based on the idea of tracking numbers but with the twist that voters do not learn their number until after voting has finished and the tracker/vote pairs have been posted to the bulletin board. This prevents the usual coercer attack on such systems: the coercer demands that the voter hand over her tracking number before the results are posted. We also provide a mix net construction that ensures that each voter gets a unique tracking number, preventing the attack of assigning the same tracker to voters likely to vote the same way. Furthermore, the construction ensures a high level of assurance that the voter receives the correct tracker while ensuring that this is deniable.

The resulting scheme provides a very direct and simple to understand mechanism for voter verification while at the same time providing receipt freeness and mitigation of coercion. The crypto is kept under the bonnet for ordinary voters, and in particular the voter verification step involves checking tracking numbers and votes in the clear. Voters do not have to handle encrypted ballots as is the case for previous E2E verifiable schemes. A further advantage appears to be that we avoid the need to audit the ballots created by the voter's device. Typically this necessitates the introduction of some kind of cut-and-chose protocol into the voting ceremony, significantly complicating the voter experience. Now, because the voter gets to check her vote in the clear we can sidestep this complication, but at the cost of a more complex dispute resolution procedure. For future research, it would be interesting to perform a usability experiment on the Selene protocol to gauge the user experience compared to other e-voting schemes.

The Selene construction can be thought of as an add-on to existing non-verifiable schemes, or indeed a conventional E2E verifiable scheme for which people want a greater degree of transparency in the verification. Indeed Selene could even be retrofitted to a cryptographic election that has been contested. Note further that an option is to run the basic Selene I scheme but if a significant level of coercion is reported before and during the vote casting period, the Selene II constructions could be dynamically added to the WBB give the higher degree of coercion resistance.

Note, Selene as presented here is intended for internet voting, but it would doubtless be straightforward to adapt it to in-person voting.

Acknowledgements. We would like to thank Sunoo Park and Bill Roscoe and many others for interesting discussions and suggestions. This work was partly supported by the FNR Luxembourg.

References

1. Adida, B., Neff, C.A.: Ballot casting assurance. In: Proceedings of the USENIX/Accurate Electronic Voting Technology Workshop, 2006 on Electronic Voting Technology Workshop (EVT 2006), Berkeley, CA, USA, p. 7. USENIX Association (2006)
2. Ben-Nun, J., Fahri, N., Llewellyn, M., Riva, B., Rosen, A., Ta-Shma, A., Wikström, D.: A new implementation of a dual (paper and cryptographic) voting system. In: 5th International Conference on Electronic Voting (EVOTE) (2012)
3. Benaloh, J.: Simple verifiable elections. In: Wallach, D.S., Rivest, R.L. (eds.) 2006 USENIX/ACCURATE Electronic Voting Technology Workshop (EVT 2006), Vancouver, BC, Canada, 1 August 2006. USENIX Association (2006)
4. Bursuc, S., Grewal, G.S., Ryan, M.D.: Trivitas: voters directly verifying votes. In: Kiayias, A., Lipmaa, H. (eds.) Vote-ID 2011. LNCS, vol. 7187, pp. 190–207. Springer, Heidelberg (2012). doi:10.1007/978-3-642-32747-6_12
5. Clarkson, M.R., Chong, S., Myers, A.C.: Civitas: a secure voting system. In: IEEE Symposium on Security and Privacy (2008)
6. Cortier, V., Fuchsbauer, G., Galindo, D.: Beleniosrf: a strongly receipt-free electronic voting scheme. IACR Cryptology ePrint Archive, 629 (2015)
7. Cramer, R., Gennaro, R., Schoenmakers, B.: A secure and optimally efficient multi-authority election scheme. In: Fumy, W. (ed.) EUROCRYPT 1997. LNCS, vol. 1233, pp. 103–118. Springer, Heidelberg (1997). doi:10.1007/3-540-69053-0_9
8. Bell, S., et al.: Star-vote: a secure, transparent, auditable, and reliable voting system. In: 2013 Electronic Voting Technology Workshop/Workshop on Trustworthy Elections (EVT/WOTE 13), Washington, D.C. USENIX Association (2013)
9. Juels, A., Catalano, D., Jakobsson, M.: Coercion-resistant electronic elections. In: Proceedings of the 2005 ACM Workshop on Privacy in the Electronic Society (WPES 2005), Alexandria, VA, USA, 7 November 2005, pp. 61–70 (2005)
10. Ramchen, K., Teague, V.: Parallel shuffling and its application to prêt à voter. In: 2010 Electronic Voting Technology Workshop/Workshop on Trustworthy Elections (EVT/WOTE 2010), Washington, D.C., USA, 9–10 August 2010 (2010)
11. Randell, B., Ryan, P.Y.A.: Voting technologies and trust. In: IEEE Symposium on Security and Privacy, pp. 50–56 (2006)

12. Rivest, R.L.: The ThreeBallot voting system. https://people.csail.mit.edu/rivest/Rivest-TheThreeBallotVotingSystem.pdf
13. Ryan, P.Y.A., Rønne, P.B., Iovino, V.: Selene: voting with transparent verifiability and coercion-mitigation. In: Clark, J., Meiklejohn, S., Ryan, P.Y.A., Wallach, D., Brenner, M., Rohloff, K. (eds.) FC 2016. LNCS, vol. 9604, pp. 176–192. Springer, Heidelberg (2016). doi:10.1007/978-3-662-53357-4_12
14. Ryan, P.Y.A.: Pretty good democracy. In: Christianson, B., Malcolm, J.A., Matyáš, V., Roe, M. (eds.) Security Protocols 2009. LNCS, vol. 7028, pp. 131–142. Springer, Heidelberg (2013). doi:10.1007/978-3-642-36213-2_16
15. Ryan, P.Y.A., Schneider, S.A.: Prêt à voter with re-encryption mixes. Technical report CS-TR-956, University of Newcastle (2006)
16. Scantegrity Team. Scantegrity. http://www.scantegrity.org/papers/whitepaper.pdf
17. Stark, P.B., Lindeman, M.: A gentle introduction to risk-limiting audits. IEEE Secur. Priv. **10**, 42–49 (2012)
18. Wagner, D., Stark, P.B.: Evidence-based elections. IEEE Secur. Priv. **10**, 33–41 (2012)

Exploring Naccache-Stern Knapsack Encryption

Éric Brier[1], Rémi Géraud[2], and David Naccache[2(✉)]

[1] Ingenico Terminals, 9 Avenue de la Gare, 26300 Alixan, France
eric.brier@ingenico.com
[2] École Normale Supérieure, 45 Rue d'Ulm, 75230 Paris Cedex 05, France
{remi.geraud,david.naccache}@ens.fr

Abstract. The Naccache-Stern public-key cryptosystem (NS) relies on the conjectured hardness of the modular multiplicative knapsack problem: Given $p, \{v_i\}, \prod v_i^{m_i} \bmod p$, find the $\{m_i\}$.

Given this scheme's algebraic structure it is interesting to systematically explore its variants and generalizations. In particular it might be useful to enhance NS with features such as semantic security, re-randomizability or an extension to higher-residues.

This paper addresses these questions and proposes several such variants.

1 Introduction

In 1997, Naccache and Stern (NS, [15]) presented a public-key cryptosystem based on the conjectured hardness of the modular multiplicative knapsack problem. This problem is defined as follows:

Let p be a modulus[1] and let $v_0, \dots, v_{n-1} \in \mathbb{Z}_p$.

$$\text{Given } p, v_0, \dots, v_{n-1}, \text{ and } \prod_{i=0}^{n-1} v_i^{m_i} \bmod p, \text{ find the } \{m_i\}.$$

Given this scheme's algebraic structure it is interesting to determine if variants and generalizations can add to NS features such as semantic security, re-randomizability or extend it to operate on higher-residues.

This paper addresses these questions and explores several such variants.

1.1 The Original Naccache-Stern Cryptosystem

The NS cryptosystem uses the following sub-algorithms:

– Setup: Pick a large prime p and a positive integer n.

---

[1] p is usually prime but nothing prevents extending the problem to composite RSA moduli.

© Springer International Publishing AG 2017
P. Farshim and E. Simion (Eds.): SecITC 2017, LNCS 10543, pp. 67–82, 2017.
https://doi.org/10.1007/978-3-319-69284-5_6

Let $\mathfrak{P} = \{p_0 = 2, \ldots, p_{n-1}\}$ be the set of the n first primes, so that

$$\prod_{i=0}^{n-1} p_i < p$$

(We leave aside a one-bit leakage dealt with in [15] — this technique applies *mutatis mutandis* to the algorithm presented in this paper).

- KeyGen: Pick a secret integer $s < p - 1$, such that $\gcd(p - 1, s) = 1$. Set

$$v_i = \sqrt[s]{p_i} \bmod p.$$

The public key is $(p, n, v_0, \ldots, v_{n-1})$. The private key is s.

- Encrypt: To encrypt an n-bit message m, compute the ciphertext c:

$$c = \prod_{i=0}^{n-1} v_i^{m_i} \bmod p$$

where m_i is the i-th bit of m.

- Decrypt: To decrypt c, compute

$$m = \sum_{i=0}^{n-1} 2^i \mu_i(c, s, p)$$

where $\mu_i(c, s, p) \in \{0, 1\}$ is the function defined by:

$$\mu_i(c, s, p) = \frac{\gcd(p_i, c^s \bmod p) - 1}{p_i - 1}.$$

To this day, NS has neither been proven secure in the usual models, nor has it been attacked. Rather, its security relies on the conjectured hardness of a multiplicative variant of the knapsack problem[2]:

Definition 1 (Multiplicative Knapsack Problem). *Given p, c, and a set $\{v_i\}$, find a binary vector x such that*

$$c = \prod_{i=0}^{n-1} v_i^{x_i} \bmod p.$$

Just as in additive knapsacks, this problem is NP-hard in general but can be solved efficiently in some situations; the secret key enabling precisely to transform the ciphertext into an easily-solvable instance.

Unlike additive knapsacks, this multiplicative knapsack doesn't lend itself to lattice reduction attacks, which completely break many additive knapsack-based cryptosystems [1,3,5,11–13].

Over the past years, several NS variants were published, these notably seek to either increase efficiency [6] or extend NS to polynomial rings [11]; to the best of our knowledge, no efficient attacks against the original NS are known.

---

[2] This can also be described as a modular variant of the "subset product" problem.

1.2   Security Notions

A cryptosystem is semantically secure, or equivalently IND-CPA-secure [9], if there is no adversary $\mathcal{A}$ capable of distinguishing between two ciphertexts of plaintexts of his choosing.

To capture this notion, $\mathcal{A}$ starts by creating two messages m_0 and m_1 and sends them to a challenger $\mathcal{C}$. $\mathcal{C}$ randomly selects one of the m_i (hereafter m_b) and encrypts it into a ciphertext c. $\mathcal{A}$ is then challenged with c and has to guess b with probability significantly higher than $1/2$.

Given a public-key cryptosystem PKC = {Setup, KeyGen, Encrypt, Decrypt}, this security notion can be formally defined by the following game:

Definition 2 (IND-CPA-Security). *The following game is played:*

- *$\mathcal{C}$ selects a secret random bit b;*
- *$\mathcal{A}$ outputs two messages m_0 and m_1;*
- *$\mathcal{C}$ sends to $\mathcal{A}$ the ciphertext $c \leftarrow$ Encrypt(m_b);*
- *$\mathcal{A}$ outputs a guess b'.*

$\mathcal{A}$ wins the game if $b' = b$. The advantage of $\mathcal{A}$ in this game is defined as:

$$\mathsf{Adv}_{\mathsf{PKC},\mathcal{A}}^{\mathsf{IND\text{-}CPA}} := \left| \Pr\left[b = b' \right] - \frac{1}{2} \right|$$

A public-key cryptosystem PKC *is* IND-CPA-secure *if* $\mathsf{Adv}_{\mathsf{PKC},\mathcal{A}}^{\mathsf{IND\text{-}CPA}}$ *is negligible for all PPT adversaries $\mathcal{A}$.*

IND-CPA-security is a very basic requirement, and in some scenarios it is desirable to have stronger security notions, capturing stronger adversaries. The strongest security notion for a public-key cryptosystem is indistinguishability under adaptive chosen ciphertext attacks, or IND-CCA2-security. IND-CCA2 is also defined in terms of a game, where $\mathcal{A}$ is furthermore given access to an encryption oracle and a decryption oracle:

Definition 3 (IND-CCA2-Security). *An adversary $\mathcal{A}$ is given access to an encryption oracle $\mathcal{O}_E$ and a decryption oracle $\mathcal{O}_D$. The following game is played:*

- *$\mathcal{C}$ selects a secret random bit b;*
- *$\mathcal{A}$ queries $\mathcal{O}_E$ and $\mathcal{O}_D$ and outputs two messages m_0 and m_1;*
- *$\mathcal{C}$ sends to $\mathcal{A}$ the ciphertext $c \leftarrow$ Encrypt(m_b);*
- *$\mathcal{A}$ queries $\mathcal{O}_E$ and $\mathcal{O}_D$ and outputs a guess b'.*

$\mathcal{A}$ wins the game if $b' = b$ and if no query to the oracles concerned m_0 nor m_1. The advantage of $\mathcal{A}$ in this game is defined as

$$\mathsf{Adv}_{\mathsf{PKC},\mathcal{A}}^{\mathsf{IND\text{-}CCA2}} := \left| \Pr\left[b = b' \right] - \frac{1}{2} \right|$$

A public-key cryptosystem PKC *is* IND-CCA2-secure *if* $\mathsf{Adv}_{\mathsf{PKC},\mathcal{A}}^{\mathsf{IND\text{-}CCA2}}$ *is negligible for all PPT adversaries $\mathcal{A}$.*

We further remind the syntax of a perfectly re-randomizable encryption scheme [4, 10, 16]. A perfectly re-randomizable encryption scheme consists in four polynomial-time algorithms (polynomial in the implicit security parameter k):

1. KeyGen: a randomized algorithm which outputs a public key pk and a corresponding private key sk.

2. Encrypt: a randomized encryption algorithm which takes a plaintext m (from a plaintext space) and a public key pk, and outputs a ciphertext c.

3. ReRand: a randomized algorithm which takes a ciphertext c and outputs another ciphertext c'; c' decrypts to the same message m as the original ciphertext c.

4. Decrypt: a deterministic decryption algorithm which takes a private key sk and a ciphertext c, and outputs either a plaintext m or an error indicator $\perp$.

In other words:
$$\{\mathsf{sk}, \mathsf{pk}\} \leftarrow \mathsf{KeyGen}(1^k)$$

$$\mathsf{Decrypt}(\mathsf{ReRand}(\mathsf{Encrypt}(m, \mathsf{pk}), \mathsf{pk}), \mathsf{sk}) = \mathsf{Decrypt}(\mathsf{Encrypt}(m, \mathsf{pk}), \mathsf{sk}) = m$$

Note that ReRand takes only a ciphertext and a public key as input, and in particular, does not require sk.

2   Higher-Residues Naccache-Stern

The deterministic nature of NS prevents it from achieving IND-CPA-security: Indeed, a given message m_0 will always produce the same ciphertext c_0, so $\mathcal{A}$ will always win the game of Definition 2.

We now describe an NS variant that is randomized. We then show how this modification guarantees semantic security, and even CCA2 security in the random oracle model, assuming the hardness of solving the multiplicative knapsack described earlier. In doing so, we must be very careful not to introduce additional structure that an adversary could leverage. To make this very visible, we decomposed the construction into three steps, each step pointing out the flaws avoided in the final construction.

2.1   Construction Step ①

Because the modified cryptosystem uses special prime moduli, algorithms Setup and KeyGen are merged into one single Setup + KeyGen algorithm[3].

- Setup + KeyGen: Pick a large prime p such that $(p-1)/2 = as$ is a factoring-resistant RSA modulus.

---

[3] Alternatively, we can regard Setup as a *pro forma* empty algorithm.

Pick a positive integer n. Let $\mathfrak{P} = \{p_0 = 2, \ldots, p_{n-1}\}$ be the set of the n first primes, so that

$$\prod_{i=0}^{n-1} p_i < p$$

Set

$$v_i = \sqrt[s]{p_i} \bmod p$$

Let g be a generator of $\mathbb{F}_p$, and $\ell = g^{2a} \bmod p$.
The public key is $(p, n, \ell, v_0, \ldots, v_{n-1})$. The private key is s.

- Encrypt: To encrypt m, pick a random integer $k \in [1, p-2]$ and compute:

$$c = \ell^k \prod_{i=0}^{n-1} v_i^{m_i} \bmod p$$

where m_i is the i-th bit of the message m.

- Decrypt: To decrypt c compute

$$m = \sum_{i=0}^{n-1} 2^i \mu_i(c, s, p).$$

To understand why decryption works we first observe that

$$(\ell^k)^s = ((g^{2a})^k)^s = g^{k(p-1)} = 1 \bmod p.$$

Hence:

$$c^s = \left(\ell^k \prod_{i=0}^{n-1} v_i^{m_i} \right)^s = (\ell^k)^s \prod_{i=0}^{n-1} p_i^{m_i} \bmod p.$$

And we are brought back to the original NS decryption process.

The Problem: The (attentive) reader could have noted at this step that because s is large and because the p_i are very few, the odds that a p_i is an s-th residue modulo p are negligible. Hence, unless p is constructed in a very particular way, key pairs simply... cannot be constructed.[4]

A solution consisting in using a specific p and is detailed in Sect. 4. The alternative consists in proceeding with ② hereafter.

2.2   Construction Step ②

The workaround will be the following: Assume that we pick a v_i at random, raise it to the power s and get some integer π:

$$\pi = v_i^s \bmod p$$

---

[4] Note that this is obviously not be an issue with the original NS scheme.

Refresh v_i until $\pi = 0 \bmod p_i$ where π is considered as an element of $\mathbb{Z}$. (In the worst case this takes p_i trials.) Letting $y_i = \pi/p_i$, we have:

$$p_i \times y_i = v_i^s \bmod p \Rightarrow p_i = y_i^{-1} \times v_i^s = u_i \times v_i^s \bmod p$$

We will now add the u_i as auxiliary public keys.

- Setup + KeyGen: Pick a large prime p such that $(p-1)/2 = as$ is a factoring-resistant RSA modulus.

  Pick a positive integer n. Let $\mathfrak{P} = \{p_0 = 2, \ldots, p_{n-1}\}$ be the set of the n first primes, so that

  $$\prod_{i=0}^{n-1} p_i < p$$

  Generate the u_i, v_i pairs as previously described so that:

  $$p_i = u_i \times v_i^s \bmod p$$

  Let g be a generator of $\mathbb{F}_p$, and $\ell = g^{2a} \bmod p$.
  The public key is $(p, n, \ell, u_0, \ldots, u_{n-1}, v_0, \ldots, v_{n-1})$. The private key is s.

- Encrypt: To encrypt m, pick a random integer $k \in [1, p-2]$ and compute:

  $$c_0 = \ell^k \prod_{i=0}^{n-1} v_i^{m_i} \bmod p \quad \text{and} \quad c_1 = \prod_{i=0}^{n-1} u_i^{m_i}$$

  where m_i is the i-th bit of the message m.

- Decrypt: To decrypt c_0, c_1 compute

  $$m = \sum_{i=0}^{n-1} 2^i \eta_i(c_0, c_1, s, p)$$

  Where

  $$\eta_i(c_0, c_1, s, p) = \frac{\gcd(p_i, c_1 \times c_0^s \bmod p) - 1}{p_i - 1}.$$

To understand why decryption works remind that $(\ell^k)^s = 1 \bmod p$ and hence

$$c_1 \times c_0^s = \prod_{i=0}^{n-1} u_i^{m_i} \left(\ell^k \prod_{i=0}^{n-1} v_i^{m_i} \right)^s = \cancel{(\ell^k)^s} \prod_{i=0}^{n-1} (u_i v_i^s)^{m_i} = \prod_{i=0}^{n-1} p_i^{m_i} \bmod p.$$

And we are brought back to the original NS decryption process.

The Problem: The (very attentive) reader could have noted that the resulting cryptosystem *does not* achieve semantic security because the construction process of c_1 is deterministic.

2.3   Construction Step ③

The workaround is the following: we provide the sender with two extra elements of $\mathbb{Z}_p$ that will allow him to blind c_0, c_1.

To that end, pick a random $\alpha \in \mathbb{Z}_p$, let $\beta \alpha^s = 1 \bmod p$ and add α, β to the public key.

The algorithms Setup+KeyGen and Decrypt remain otherwise unchanged but Encrypt now becomes:

- Encrypt: To encrypt m, pick a random integer $k \in [1, p-2]$ and compute:

$$c_0 = \alpha^k \prod_{i=0}^{n-1} v_i^{m_i} \bmod p \quad \text{and} \quad c_1 = \beta^k \prod_{i=0}^{n-1} u_i^{m_i}.$$

To understand why decryption works we note that (modulo p):

$$c_1 \times c_0^s = \beta^k \prod_{i=0}^{n-1} u_i^{m_i} \left(\alpha^k \prod_{i=0}^{n-1} v_i^{m_i} \right)^s = (\beta \alpha^s)^k \prod_{i=0}^{n-1} (u_i v_i^s)^{m_i} = \prod_{i=0}^{n-1} p_i^{m_i}.$$

And we are brought back to the original NS decryption process.

3   Security

3.1   Semantic Security

The modified scheme's security essentially relies on blinding an NS ciphertext using a multiplicative factor $\ell^k = g^{2ka} \bmod p$, which belongs to the subgroup of $\mathbb{Z}_p$ of order b.

Lemma 1. *Under the subgroup hiding assumption in $\mathbb{Z}_p$, the scheme described in Sect. 2.1 is* IND-CPA-*secure.*

Recall that the subgroup-hiding assumption [2] states that the uniform distribution over $\mathbb{Z}_p$ is indistinguishable from the uniform distribution over one of its subgroups.

Proof. Assume that $\mathcal{A}(\mathsf{pk})$ wins the IND-CPA game with non-negligible advantage. Then in particular $\mathcal{A}(\mathsf{pk})$ has non-negligible advantage in the "real-or-random" game

$$\mathsf{Adv}_{\mathcal{A}}^{\mathsf{R/R}} := \Pr[\mathcal{A}^{\mathcal{E}_{\mathsf{pk}}}(\mathsf{pk}) = 1] - \Pr[\mathcal{A}^{\mathcal{O}}(\mathsf{pk}) = 1]$$

where $\mathcal{E}_{\mathsf{pk}}$ is an encryption oracle and $\mathcal{O}$ is a random oracle. We define $\mathcal{B}(\mathsf{pk}, \gamma)$ as follows:

- Let $\mathcal{E}_{\mathcal{B}}(m) = \gamma \prod_{i=0}^{n-1} v_i^{m_i} \bmod p$;
- $\mathcal{B}(\mathsf{pk}, \gamma)$ returns the same result as $\mathcal{A}^{\mathcal{E}_{\mathcal{B}}}(\mathsf{pk})$

The scenario $\mathcal{B}(\mathsf{pk}, \gamma = g^{2au})$ yields $\mathcal{E}_\mathcal{B} = \mathcal{E}_\mathsf{pk}$. The scenario $\mathcal{B}(\mathsf{pk}, \gamma = g^u)$ for random u gives a ciphertext that is a uniform value, and therefore behaves as a perfect simulator of a random oracle, i.e. $\mathcal{E}_\mathcal{B} = \mathcal{O}$. Hence if $\mathcal{A}$ is an efficient adversary against our scheme, then $\mathcal{B}$ is an efficient solver for the subgroup-hiding problem. □

Note that this part of the argument does not fundamentally rely on the original NS being secure — indeed, we may consider an encryption scheme that produces ciphertexts of the form $c = x^k m$. Decryption for such a cryptosystem would be tricky, as $c^b = m^b$ and there are b possible roots. That is why using NS is useful, as we do not have decryption ambiguity issues.

As we pointed out, the construction of Sect. 2.2 is not semantically secure: indeed, c_1 is generated deterministically from m. This is addressed in Sect. 2.3 by introducing two numbers α and β. Using a similar argument as in Lemma 1, we have

Lemma 2. *Under the DDH assumption in $\mathbb{Z}_p$, and assuming that factoring $(p-1)/2$ is infeasible, the scheme described in Sect. 2.3 is* IND-CPA-*secure.*

Note that these hypotheses can be simultaneously satisfied.

3.2   CCA2 Security

Even more interesting is the case for security against adaptive chosen-ciphertext attacks (IND-CCA2) [7,8].

The original NS is naturally not IND-CCA2; nor is in fact the "Step ①" variant discussed above: indeed it is possible to re-randomise a ciphertext, which immediately gives a way to win the IND-CCA2 game.

To remedy this, we leverage the fact that upon successful decryption, we can recover the randomness ℓ^k. The idea is to choose k in some way that depends on m_i. If k is a deterministic function of m_i only however, randomisation is lost. Therefore we suggest the following variant, at the cost of some bandwidth:

- Instead of m, we encrypt a message $m\|r$ where r is a random string.
- Let $k \leftarrow H(m\|r)$ where H is a cryptographic hash function, and use this value of k instead of choosing it randomly in Encrypt.
- Modify Decrypt to recover ℓ^k (or α^k and β^k). Upon successfully recovering $(m\|r)$, extract r, and check that ℓ^k (resp. α^k and β^k) correspond to the correct value of k — otherwise it outputs $\bot$.

This approach guarantees IND-CCA2 in the random oracle model; this can be captured as a series of games:

- *Game 0*: This is the IND-CCA2 game against our scheme (① or ③), instantiated with some hash function H.
- *Game 1*: This game differs from *Game 0* in replacing H by a random oracle $\mathcal{O}$. In the random oracle model, this game is computationally indistinguishable from *Game 0*.

– *Game 2*: This game differs from *Game 1* by the fact that the ciphertext is replaced by an uniformly-sampled random element of the ciphertext space. The results on IND-CPA security tell us that this game is computationally indistinguishable from *Game 1* (under their respective hypotheses).

4   Generating Strong Pseudo-Primes in Several Bases

We now backtrack and turn our attention to generating specific moduli allowing to implement securely the "①" scheme of Sect. 2.1. This boils down to describing how to efficiently generate strong pseudo-prime numbers. In this section, we denote N the sought-after modulus.

Using quadratic reciprocity, we first introduce an algorithm generating numbers passing Fermat's test. Then we leverage quartic reciprocity to generate numbers passing Miller-Rabin's test. The pseudoprimes we need must be strong over several bases, and complexity is polynomial in the size of the product of these bases.

4.1   Primality Tests

A base-A Fermat primality test consists in checking that $A^B \equiv A \bmod B$. Every prime passes this test for all bases A. There are however composite numbers, known as Carmichael numbers, that also pass this test in all bases. For instance, $1729 = 7 \cdot 13 \cdot 19$ is such a number. There are an infinity of Carmichael numbers. The Miller-Rabin primality test also relies on Fermat's little theorem. Let $B-1 = 2^e m$ with m odd. An integer B passes the Miller-Rabin test if $A^m \equiv 1 \bmod B$ or if there exists an $i \leq e - 1$ such that $A^{2^i m} \equiv -1 \bmod B$.

Definition 4 (Strong pseudo-prime). *A number that passes the Miller-Rabin test is said strongly pseudo-prime in base A.*

An interesting theorem [14, Proposition 2][17] states that a composite number can only be strongly pseudo-prime for a quarter of the possible bases.

4.2   Constructing Pseudo-Primes

When p and $2p - 1$ are prime, Fermat's test amounts to the computing of a Jacobi symbol. Indeed,

Theorem 1. *Let p be a prime such that $q = 2p - 1$ is also prime. Let $A \in \mathrm{QR}_q$. Then $B = pq$ passes Fermat's test in base A.*

Proof.

$$A^B \equiv (A^p)^q \equiv A^q \equiv A^{2(p-1)+1} \equiv A \bmod p$$

$$A^B \equiv (A^q)^p \equiv A^p \equiv A^{(q-1)/2+1} \equiv A \left(\frac{A}{q} \right) \equiv A \bmod q$$

By the Chinese remainder theorem, we find that $A^B \equiv A \bmod B$.   □

From Gauss' quadratic reciprocity theorem, if $q \equiv 1 \bmod 4$ we can take $q \equiv 1 \bmod A$ which guarantees that $A \in \mathrm{QR}_q$. To make 2 a quadratic residue modulo q we must have $q \equiv \pm 1 \bmod 8$. It is therefore easy to construct numbers that pass Fermat's test in a prescribed list of bases.

4.3   Constructing Strong Pseudo-Primes

In this section we seek to generate numbers that are strongly pseudo-prime in base η, where η is prime. Let p denote a prime number such that $q = 2p - 1$ is also prime, and $N = pq$. We have the following equations:

$$N - 1 \equiv 0 \bmod p - 1$$
$$N - 1 \equiv \frac{q-1}{2} \bmod q - 1$$
$$\frac{n-1}{2} \equiv \frac{p-1}{2} \bmod p - 1$$
$$\frac{n-1}{2} \equiv 3\frac{q-1}{4} \bmod q - 1$$

From there on, we will use the notation $\left(\frac{\cdot}{\cdot}\right)_4$ to denote the quartic residue symbol.

Theorem 2. *Let p be a prime such that $q = 2p - 1 \equiv 1 \bmod 8$ is also prime. Let A be an integer such that*

$$\left(\frac{A}{p}\right) = -1, \qquad \left(\frac{A}{q}\right) = +1, \quad and \quad \left(\frac{A}{q}\right)_4 = -1.$$

Then $N = pq$ passes the Miller-Rabin test in base A.

Proof. Note that if $A^{(N-1)/2} \equiv -1 \bmod N$, then n passes the Miller-Rabin test in base a. It then suffices to compute this quantity modulo p and q respectively:

$$A^{(N-1)/2} \equiv A^{(p-1)/2} \equiv \left(\frac{A}{p}\right) \equiv -1 \bmod p$$
$$A^{(N-1)/2} \equiv A^{3(q-1)/4} \equiv \left(\frac{A}{p}\right)_4^3 \equiv -1 \bmod q.$$

$\square$

Bases $\eta > 5$. Let $\eta \geq 7$ be a prime number. We consider here the case $p \equiv 5 \bmod 8$, i.e. $q \equiv 9 \bmod 16$. We will leverage the following classical result:

Theorem 3. *Let q be a prime number, $q = A^2 + B^2 \equiv 1 \bmod 8$ with B even. Let η be a prime number such that $(p/\eta) = 1$, then*

$$\left(\frac{\eta}{q}\right)_4 = 1 \Leftrightarrow \begin{cases} \eta \mid B & , \ or \\ \eta \mid A \ and \ \left(\frac{2}{\eta}\right) = 1 & , \ or \\ A \equiv \mu B \ where \ \mu^2 + 1 \equiv \lambda^2 \bmod \eta \ and \ \left(\frac{\lambda(\lambda+1)}{\eta}\right) = 1 \end{cases}$$

We will also need the following easy lemmata:

Lemma 3. *Let $\eta \geq 7$ be a prime number, there is at least an integer Λ such that*

$$\left(\frac{\Lambda}{\eta}\right) = \left(\frac{2-\Lambda}{\eta}\right) = -1.$$

Proof. Let

$$s_1 = \#\left\{i \in \mathbb{F}_\eta, \left(\frac{i}{\eta}\right) = +1, \left(\frac{2-i}{\eta}\right) = +1, \right\}$$

$$s_2 = \#\left\{i \in \mathbb{F}_\eta, \left(\frac{i}{\eta}\right) = +1, \left(\frac{2-i}{\eta}\right) = -1, \right\}$$

$$s_3 = \#\left\{i \in \mathbb{F}_\eta, \left(\frac{i}{\eta}\right) = -1, \left(\frac{2-i}{\eta}\right) = +1, \right\}$$

$$s_4 = \#\left\{i \in \mathbb{F}_\eta, \left(\frac{i}{\eta}\right) = -1, \left(\frac{2-i}{\eta}\right) = -1, \right\}.$$

Then it is clear that $s_1 + s_2 + s_3 + s_4 = \eta - 2$. The quantity $s_1 + s_2$ corresponds to the number of quadratic residues modulo η, except maybe 2. Therefore,

$$s_1 + s_2 = \frac{\eta - \left(\frac{2}{\eta}\right)}{2} - 1.$$

By symmetry between i and $2 - i$, we have $s_2 = s_3$. We also have

$$s_2 + s_3 = \#\left\{i \in \mathbb{F}_\eta, \left(\frac{i(2-i)}{\eta}\right) = -1\right\}$$

$$= \#\left\{i \in \mathbb{F}_\eta^*, \left(\frac{2/i - 1}{\eta}\right) = -1\right\}$$

$$= \#\left\{u \in \mathbb{F}_\eta, u \neq -1, \left(\frac{u}{\eta}\right) = -1\right\}$$

$$= \frac{\eta + \left(\frac{-1}{\eta}\right)}{2} - 1.$$

From that we get the value of s_4:

$$s_4 = \frac{\eta + 2\left(\frac{2}{\eta}\right) - \left(\frac{-1}{\eta}\right) - 2}{4}.$$

Therefore, for every $\eta \geq 7$, $s_4 > 0$. $\square$

Choosing such an i, we denote λ the integer such that $i = 1 + 1/\lambda \bmod \eta$. Then,

$$\left(\frac{1 + 1/\lambda}{\eta}\right) = \left(\frac{1 - 1/\lambda}{\eta}\right) = -1$$

$$\left(\frac{(\lambda + 1)\lambda}{\eta}\right) = \left(\frac{(\lambda - 1)\lambda}{\eta}\right) = -1$$

$$\left(\frac{\lambda^2 - 1}{\eta}\right) = \left(\frac{(\lambda + 1)\lambda}{\eta}\right)\left(\frac{(\lambda - 1)\lambda}{\eta}\right) = 1.$$

Let μ be such that $\mu^2 + 1 = \lambda^2$. We can thus construct λ and μ so that the third possibility of Theorem 3 is never satisfied.

Lemma 4. *Let $\eta \geq 7$ be a prime number, there is at least an integer x such that $(x/\eta) = -1$ and $(2x - 1/\eta) = +1$.*

Proof. As for the previous lemma, we show that there are $\frac{1}{4}\left(\eta + 2\left(\frac{2}{\eta}\right) - \left(\frac{-1}{\eta}\right) - 2\right)$ such values of x, which strictly positive for $\eta \geq 7$. $\square$

For such an x, we write $y = 2x - 1 = z^2 \bmod \eta$, $A_\eta = z/\lambda \bmod \eta$, and $B_\eta = A_\eta\mu$. We then have

$$A_\eta^2 + B_\eta^2 = (1 + \mu^2)A_\eta^2 = \lambda^2 A_\eta^2 = z^2 = y \bmod \eta$$

If $q = A^2 + B^2 \equiv 1 \bmod 8$ is prime, with B even, $A \equiv A_\eta \bmod \eta$, and $B \equiv B_\eta \bmod \eta$, then we see that the conditions of Theorem 3 are not satisfied, hence $(\eta/q)_4 = -1$. Furthermore, $q \equiv y \bmod \eta$ so that $(\eta/q) = +1$. If we assume that $p = (q + 1)/2$ is prime, and that $p \equiv 5 \bmod 8$, then the conditions of Theorem 2 are satisfied. Indeed, $p \equiv x \bmod \eta$ so that $(\eta/p) = (x/\eta) = -1$. Thus we generated a pseudo-prime in base η.

All in all, the results from this section are captured by the following theorem.

Theorem 4. *Let $\eta \geq 7$ be a prime number. There are integers A_η, A_η such that $N = pq$ is strongly pseudo-prime in base η, provided that*

$$\begin{cases} q & = A^2 + B^2 \\ B & \text{is even} \\ A & \equiv A_\eta \bmod \eta \\ B & \equiv B_\eta \bmod \eta \\ q & \equiv 9 \bmod 16 \\ p & = (q - 1)/2 \\ q & \text{is prime} \\ p & \text{is prime} \end{cases}$$

Base $\eta = 2$. In that case the following theorem applies.

Theorem 5. *The integer $N = pq$ is strongly pseudo-prime in base 2 provided that*

$$\begin{cases} q & = A^2 + B^2 \\ A & \equiv 3 \bmod 8 \\ B & \equiv 4 \bmod 8 \\ p & = (q-1)/2 \\ q & \text{is prime} \\ p & \text{is prime} \end{cases}$$

Proof. From the conditions of theorem 5, $q \equiv 9 \bmod 16$ and $q \equiv 5 \bmod 8$, which proves that 2 is a square modulo q and not modulo p, as it is not of the form $\alpha^2 + 64\beta^2$. □

Bases $\eta = 3$ and $\eta = 5$. In both cases, we cannot find p and q such that the base is a square modulo q and not modulo p. As we will see in the next section this is not too much of a problem in practice. We can in any case ensure that the base is a quartic residue modulo q, using for instance the following choices:

$$A_3 = 1, \qquad B_3 = 0,$$
$$A_5 = 1, \qquad B_5 = 0.$$

4.4 Combining Bases

Consider a set $\mathfrak{P}$ of prime numbers, which will be used as bases. For each $\eta \in \mathfrak{P}$, we construct a_η, b_η as described in the previous section, using either the general construction (for $\eta \geq 7$) or the specific constructions (for $\eta = 2, 3, 5$). Then we invoke the Chinese remainder theorem, to get three integers $a_\mathfrak{P}$, $b_\mathfrak{P}$, and $m_\mathfrak{P}$ such that $N = pq$ is strongly pseudo-prime in all bases of $\mathfrak{P}$ (except maybe 3 and 5), provided that

$$\begin{cases} q & = A^2 + B^2 \\ B & \text{is even} \\ A & \equiv A_\mathfrak{P} \bmod m_\mathfrak{P} \\ B & \equiv B_\mathfrak{P} \bmod m_\mathfrak{P} \\ q & \equiv 9 \bmod 16 \\ p & = (q-1)/2 \\ q & \text{is prime} \\ p & \text{is prime} \end{cases}$$

In fact, running the algorithm several times eventually yields an integer N that is also strongly pseudo-prime in bases 3 and 5.

4.5 Numerical Example

Consider $\mathfrak{P} = \{p_1 = 2, \ldots, p_{46}\}$ the set of all primes smaller than 200. We get:

$$
\begin{aligned}
A_{\mathfrak{P}} = \ & 24095104664133668361029398948772093859437 0 \\
& 0042913129394126042848260031865186440501 1 \\
B_{\mathfrak{P}} = \ & 24500136562064551260427880199750830122812 \\
& 89375458232594038192481071092303905088660 \\
m_{\mathfrak{P}} = \ & 31199688166733846212996796425319255506751 9 \\
& 87159614203780372129899474046144658803240
\end{aligned}
$$

From these we get the following number N, which is strongly pseudo-prime over all the bases in $\mathfrak{P}$:

$$
\begin{aligned}
p = \ & 29161850666397983648507555237542534127102 9 \\
& 35727619494034905899381284476833930749393 8 \\
& 12764659482181700902524129015037164276859 7 \\
& 76144331858469203988770750118933523764312 1 \\
& 80942186641722156221 \\
q = \ & 58323701332795967297015110475085068254205 8 \\
& 71455238988069811798762568953667861498787 6 \\
& 25529318964363401805048258030074328553719 5 \\
& 52288663716938407977541500237867047528624 3 \\
& 61884373283444312441 \\
N = \ & 17008270685785930460134654204088049186996 4 \\
& 78613827323514836026400701165992713709380 9 \\
& 42510806917357993787977335822184994450664 6 \\
& 59888785836135840319726564065098289305232 8 \\
& 56031565088228413420696658370388670205884 \\
& 47417990839513625631031172048540249389031 2 \\
& 41584596856378126949009288986603857918379 1 \\
& 39501994817399415095992110561507861273999 9 \\
& 52621422448462073244786658072173358454 61
\end{aligned}
$$

This N can hence be used as the missing modulus needed to instantiate a "Step ①" NS variant.

5 Extensions

5.1 Using Composite Moduli

In the ②/③ variants of our scheme, one might be tempted to replace p *itself* by an RSA modulus n, where $\phi(n) = 2ab$. Indeed, the original NS construction allows for such a choice.

Doing so, however, would immediately leak information about the factorisation of n: Indeed, $\gcd(g^a - 1, n) = p$.

There is a workaround: First we choose p and q so that $(p-1)/2$ and $(q-1)/2$ are RSA moduli, i.e. $p - 1 = 2s_1 s_2$ and $q - 1 = 2r_1 r_2$, with large s_1, s_2, r_1, r_2. Then we set $n = pq$, $a = s_1 r_1$, and $b = 2s_2 r_2$. Therefore $\phi(n) = 2ab$ as before, but the GCD attack mentioned above does not apply, and the modified ②/③ Naccache-Stern cryptosystem works.

5.2 Bandwidth Improvements

The idea described in this paper is fully compatible with the modifications introduced in [6] to improve encryption bandwidth.

But there is even more: An interesting observation is that, upon decryption, it is possible to recover both the message m and the whitening x^k. This is unlike most randomized encryption schemes, where the random nonce is lost. Thus we may contemplate storing some information in k, thereby augmenting somewhat the total information contained in a ciphertext. Alternatively, x^k may also be used as key material if NS is used (in a hybrid mode) as a key transfer mechanism.

For instance, given a message $m = m_1 \| m_2$, we may encrypt $m_1 \| k$ using the blinding m_2^k with odd k. Upon decryption, one recovers k, and computes the k-th root of the blinding factor m_2^k — such a root is unique with overwhelming probability — thereby reconstructing the whole message.

One nontrivial research direction is to provide, in the message m, *hints* that make solving the discrete log modulo p easier and thereby embed directly information in k.

References

1. Adleman, L.M.: On breaking the iterated Merkle-Hellman public-key cryptosystem. In: Chaum, D., Rivest, R.L., Sherman, A.T. (eds.) Advances in Cryptology - CRYPTO 1982, pp. 303–308. Plenum Press, New York (1982)
2. Boneh, D., Goh, E.-J., Nissim, K.: Evaluating 2-DNF formulas on ciphertexts. In: Kilian, J. (ed.) TCC 2005. LNCS, vol. 3378, pp. 325–341. Springer, Heidelberg (2005). doi:10.1007/978-3-540-30576-7_18
3. Brickell, E.F.: Breaking iterated Knapsacks. In: Blakley, G.R., Chaum, D. (eds.) CRYPTO 1984. LNCS, vol. 196, pp. 342–358. Springer, Heidelberg (1985). doi:10.1007/3-540-39568-7_27
4. Canetti, R., Krawczyk, H., Nielsen, J.B.: Relaxing chosen-ciphertext security. In: Boneh, D. (ed.) CRYPTO 2003. LNCS, vol. 2729, pp. 565–582. Springer, Heidelberg (2003). doi:10.1007/978-3-540-45146-4_33

5. Chee, Y.M., Joux, A., Stern, J.: The cryptanalysis of a new public-key cryptosystem based on modular Knapsacks. In: Feigenbaum, J. (ed.) CRYPTO 1991. LNCS, vol. 576, pp. 204–212. Springer, Heidelberg (1992). doi:10.1007/3-540-46766-1_15

6. Chevallier-Mames, B., Naccache, D., Stern, J.: Linear bandwidth Naccache-Stern encryption. In: Ostrovsky, R., De Prisco, R., Visconti, I. (eds.) SCN 2008. LNCS, vol. 5229, pp. 327–339. Springer, Heidelberg (2008). doi:10.1007/978-3-540-85855-3_22

7. Cramer, R., Shoup, V.: A practical public key cryptosystem provably secure against adaptive chosen ciphertext attack. In: Krawczyk, H. (ed.) CRYPTO 1998. LNCS, vol. 1462, pp. 13–25. Springer, Heidelberg (1998). doi:10.1007/BFb0055717

8. Fujisaki, E., Okamoto, T., Pointcheval, D., Stern, J.: RSA-OAEP is secure under the RSA assumption. J. Cryptology 17(2), 81–104 (2004)

9. Goldwasser, S., Micali, S.: Probabilistic encryption and how to play mental poker keeping secret all partial information. In: Lewis, H.R., Simons, B.B., Burkhard, W.A., Landweber, L.H. (eds.) Proceedings of the 14th Annual ACM Symposium on Theory of Computing, 5–7 May 1982, San Francisco, California, USA, pp. 365–377. ACM (1982)

10. Groth, J.: Rerandomizable and replayable adaptive chosen ciphertext attack secure cryptosystems. In: Naor, M. (ed.) TCC 2004. LNCS, vol. 2951, pp. 152–170. Springer, Heidelberg (2004). doi:10.1007/978-3-540-24638-1_9

11. Herold, G., Meurer, A.: New attacks for Knapsack based cryptosystems. In: Visconti, I., De Prisco, R. (eds.) SCN 2012. LNCS, vol. 7485, pp. 326–342. Springer, Heidelberg (2012). doi:10.1007/978-3-642-32928-9_18

12. Joux, A., Stern, J.: Cryptanalysis of another Knapsack cryptosystem. In: Imai, H., Rivest, R.L., Matsumoto, T. (eds.) ASIACRYPT 1991. LNCS, vol. 739, pp. 470–476. Springer, Heidelberg (1993). doi:10.1007/3-540-57332-1_40

13. Lenstra, H.W.: On the Chor-Rivest Knapsack cryptosystem. J. Cryptology 3(3), 149–155 (1991)

14. Monier, L.: Evaluation and comparison of two efficient probabilistic primality testing algorithms. Theoret. Comput. Sci. 12(1), 97–108 (1980)

15. Naccache, D., Stern, J.: A new public-key cryptosystem. In: Fumy, W. (ed.) EUROCRYPT 1997. LNCS, vol. 1233, pp. 27–36. Springer, Heidelberg (1997). doi:10.1007/3-540-69053-0_3

16. Prabhakaran, M., Rosulek, M.: Rerandomizable RCCA encryption. In: Menezes, A. (ed.) CRYPTO 2007. LNCS, vol. 4622, pp. 517–534. Springer, Heidelberg (2007). doi:10.1007/978-3-540-74143-5_29

17. Rabin, M.O.: Probabilistic algorithm for testing primality. J. Number Theory 12(1), 128–138 (1980)

Proximity Assurances Based on Natural and Artificial Ambient Environments

Iakovos Gurulian[1]([⊠]), Konstantinos Markantonakis[1], Carlton Shepherd[1],
Eibe Frank[2], and Raja Naeem Akram[1]

[1] Information Security Group Smart Card Centre, Royal Holloway,
University of London, Egham, UK
{Iakovos.Gurulian.2014,k.markantonakis,Carlton.Shepherd.2014,
r.n.akram}@rhul.ac.uk
[2] Department of Computer Science, University of Waikato, Hamilton, New Zealand
eibe@waikato.ac.nz

Abstract. Relay attacks are passive man-in-the-middle attacks that aim to extend the physical distance of devices involved in a transaction beyond their operating environment. In the field of smart cards, distance bounding protocols have been proposed in order to counter relay attacks. For smartphones, meanwhile, the natural ambient environment surrounding the devices has been proposed as a potential Proximity and Relay-Attack Detection (PRAD) mechanism. These proposals, however, are not compliant with industry-imposed constraints that stipulate maximum transaction completion times, e.g. 500 ms for EMV contactless transactions. We evaluated the effectiveness of 17 ambient sensors that are widely-available in modern smartphones as a PRAD method for time-restricted contactless transactions. In our work, both similarity- and machine learning-based analyses demonstrated limited effectiveness of natural ambient sensing as a PRAD mechanism under the operating requirements for proximity and transaction duration specified by EMV and ITSO. To address this, we propose the generation of an Artificial Ambient Environment (AAE) as a robust alternative for an effective PRAD. The use of infrared light as a potential PRAD mechanism is evaluated, and our results indicate a high success rate while remaining compliant with industry requirements.

Keywords: Mobile payments · Relay attacks · Ambient environment sensing · Contactless · Experimental analysis

1 Introduction

Today, a wide variety of application environments exist that demand proximity of a user with a physical terminal, as well as high throughput, i.e. maximising the number of transactions per unit time. Both smart card-based payments and transport-related transactions are major examples of such applications in everyday life. These particular services are governed by industry-accepted standards,

© Springer International Publishing AG 2017
P. Farshim and E. Simion (Eds.): SecITC 2017, LNCS 10543, pp. 83–103, 2017.
https://doi.org/10.1007/978-3-319-69284-5_7

such as the EMV specifications for card and mobile contactless payments. Under EMV, contactless transactions should complete within 500 ms [2–4]. Similarly, transport-related transactions should complete between 300 and 500 ms [1]. In addition to these, other applications exist that depend on proximity and transaction time, particularly in the realm of the Internet of Things (IoT), such as taking medical equipment inventories in operating theatres. The domain of sensor networks is another closely-related area where communication time and the proximity of sensors can be of paramount importance.

In this paper, we examine the problem of proximity detection in applications with restricted time-frames. Specifically, we focus on applications that are deployed traditionally as contactless smart cards but are gradually migrating to mobile phones using Near-Field Communication (NFC). During an NFC-based mobile contactless transaction, a mobile handset is brought into the radio range (<3 cm) of a payment terminal through which a dialogue is initiated. NFC, however, has no provisions to ascertain whether the device is genuinely in proximity to the terminal, which makes them susceptible to relay attacks.

In a relay attack [8,9,38], the aim of the malicious actor is to extend the physical distance of the communication channel between the victim's mobile phone and the transaction terminal – relaying each message across this extended distance. The attacker extends this distance using equipment that masquerades as legitimate devices to both the terminal and victim device, as shown in Fig. 1. The attacker has the potential to gain access to services using the victim's account if messages are relayed successfully without detection. At present, additional security mechanisms, like fingerprint scanning and Personal Identity Number (PIN) entry, may also be required in order to perform a contactless mobile transaction for a payment, transport ticketing, and similar services. However, even the use of PINs and biometrics cannot always prevent relay attacks (see the Mafia fraud attack [7]).

| Victim's | Emulating | | Emulating | Payment |
| Phone | Terminal | | Victim's Phone | Terminal |

Fig. 1. Overview of a relay attack [31].

In recent years, a deluge of Proximity/Relay Attack Detection (PRAD) mechanisms have been proposed that rely on collecting information regarding the ambient environment surrounding the transaction instrument and terminal. Such proposals collect data using the sensors in modern mobile devices – such as temperature, motion and position sensors – which is compared for similarity to assure that the transaction devices were genuinely in proximity. In this work, we present an empirical evaluation of the claim that *ambient sensing on mobile devices is an effective PRAD method under the time conditions stipulated by industry*. We present an extended study before proposing the utility of an artificially generated ambient environment as an alternative PRAD mechanism.

2 Natural Ambient Environment Sensing

In this section, we discuss a number of generic deployment models for deploying proximity- and transaction time-sensitive applications using ambient sensing. Next, we discuss related work before evaluating the claim that ambient sensors are an effective PRAD mechanism under the real-world constraints imposed by industry requirements, i.e. by EMV and ITSO.

2.1 Ambient Sensors in Conventional Transactions

For contactless smart cards, relay attacks can be countered using distance bounding protocols [27] and variants of such [18]. This is still an active research domain, with new attacks and countermeasures emerging [5,7,17]. At the current state of the art, however, these are not easily transferable to NFC-enabled phones, due to their high sensitivity to time delays [6,16,35]. Alternative methods have been proposed to provide proximity detection, most of which use environmental and motion sensors present on modern mobile handsets [16,23,32,34,36,37]. In Sect. 2.2, we discuss how ambient sensors have been proposed to counter relay attacks in NFC-based mobile contactless transactions.

An ambient sensor measures a particular environmental property of its immediate surroundings, such as temperature, light, humidity and sound; a wealth of such sensors are deployed in modern smartphones and tablets. In Fig. 2, we illustrate a generic approach for deploying ambient sensing as a proximity detection mechanism for mobile payments, with the following variations:

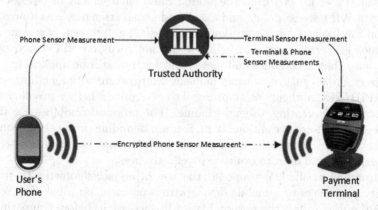

Fig. 2. Generic deployment of mobile sensing for proximity detection [31].

1. **Independent Reporting.** Both the smartphone and payment terminal collect sensor measurements independently of each other and transmit these to a trusted authority (depicted as solid lines in Fig. 2). The authority compares the sensor measurements, based on some predefined comparison algorithm with set margins of error (threshold), and decides whether the two devices are within proximity to each another.
2. **Payment Terminal Dependent Reporting.** This set-up involves the smartphone encrypting the sensor measurements with a shared key between smartphone and trusted authority, and transmitting the encrypted message to the payment terminal (shown as double-dot-dash lines in Fig. 2). The payment terminal sends its own measurements and the smartphone's to the trusted authority for comparison.
3. **Payment Terminal (Localised) Evaluation.** The smartphone transmits its measurement to the payment terminal, which compares it with its own measurements locally; the payment terminal then decides whether the smartphone is in proximity.

2.2   Related Work

We identify and summarise key pieces of related work that have suggested using natural ambient sensing as a PRAD mechanism.

Ma et al. [23] explored the use of GPS (Global Positioning System) location data for determining the proximity of two mobile phones. A ten-second recording window was used in which GPS data was collected every second, which was subsequently compared across various devices. The work reported a high success rate in identifying co-located devices.

Halevi et al. [16] demonstrated the use of ambient sound and light for proximity detection. The authors analysed sensor measurements – collected for 2 and 30 s duration for light and audio respectively – using a range of similarity comparison metrics. Extensive experiments were performed in different physical locations with a high success rate in detecting proximate devices.

Varshavsky et al. [37] used the shared radio environment of devices – the presence of WiFi access points and associated signal strengths – as a proximity detection mechanism for secure device pairing. The approach was considered to produce low error rates and, while it did not focus on NFC-based mobile transactions, their techniques and methodologies may still be applicable.

Urien et al. [36] suggested using ambient temperature with an elliptic curve-based RFID/NFC authentication protocol to determine whether two devices are co-located before creating a secure channel. The proposal combines the timing channels in RFID, used traditionally in distance bounding protocols, in conjunction with ambient temperature. The work, however, was not implemented and has no experimental data to evaluate its effectiveness.

Mehrnezhad et al. [25] proposed the use of an accelerometer to provide proximity assurances of a mobile device with a payment terminal. The scheme requires the user to tap the terminal twice in succession, before comparing the sensor data from the device and the terminal for similarity. It is difficult to deduce the total time it took to complete a transaction entirely, but the authors provide a recording time range of 0.6–1.5 s.

Truong et al. [34] evaluated four different sensors using a recording duration of 10–120 s. While the results were positive, the long sampling duration renders it unsuitable for NFC-based mobile transactions.

Additional work by Jin et al. [20] showed that a smartphone's magnetometer can be used to establish proximity assurance. This approach requires more than 500 ms; the authors do not claim that magnetometers can provide an effective relay attack detection mechanism.

Shrestha et al. [32] used a number of ambient sensors within specialised hardware, known as Sensordrone, for proximity detection. The work was not evaluated using the ambient sensors available on commodity handsets, did not provide the sampling duration, and states that data from each sensor was collected for a few seconds.

Table 1. Related work in sensor-based PRAD mechanisms.

Paper	Sensor(s) used	Sample duration	Contactless suitability
Ma et al. [23]	GPS	10 s	Unlikely
Halevi et al. [16]	Audio	30 s	Unlikely
	Light	2 s	More Likely
Varshavsky et al. [37]	WiFi (Radio waves)	1 s	More Likely
Urien et al. [36]	Temperature	N/A	–
Mehrnezhad et al. [25]	Accelerometer	0.6 to 1.5 s	More Likely
Truong et al. [34]	GPS raw data	120 s	Unlikely
	WiFi	30 s	Unlikely
	Ambient audio	10 s	Unlikely
	Bluetooth	12 s	Unlikely
Shrestha et al. [32]	Temperature (T)	Few seconds	Unlikely
	Precision Gas (G)	Few seconds	Unlikely
	Humidity (H)	Few seconds	Unlikely
	Altitude (A)	Few seconds	Unlikely
	HA	Few seconds	Unlikely
	HGA	Few seconds	Unlikely
	THGA	Few seconds	Unlikely

Table 1 summarises past work, using sensor sampling durations to determine their suitability for NFC-based mobile phone transactions in banking and transportation. 'Unlikely' proposals have sample durations so large that they may not be adequate for mobile-based services that substitute contactless cards, while those with reasonably short durations are labelled 'More Likely'. However, even schemes denoted as 'More Likely' may not be suitable as no proposal is evaluated under the time constraints stipulated by the banking and transport sectors. In these sectors, the goal is to serve people as quickly as possible to maximise customer throughput, as alluded to in Sect. 1; time is critical in determining whether a transaction is successful and, indeed, permitted. Here, an optimal transaction duration is 500 ms rather than seconds.

Our initial study (Shepherd et al. [31]) questioned the effectiveness of ambient sensing as a proximity detection mechanism under short time frames (500 ms) – illustrating that numerous sensors available via the Android platform perform poorly within an operating distance of <3 cm and transaction-duration of <500 ms. Both threshold- and machine learning-based analyses were employed using sensor data collected from mock transactions in the field. Similar results were also exhibited by further experimentations (Haken et al. [15]) using sensors on the Apple iOS platform. Our third analysis (Gurulian et al. [14]) selected seven of the best-performing sensors from our first study [31]. In this study, sensor data from genuine and relay transactions was collected from an emulated relay attack set-up, with the goal of determining whether data from relayed transactions can be distinguished from legitimate ones. In the following sections, we reproduce the results from these initial studies along with additional analyses conducted post-publication. Note that the focus of this work is on conventional transactions that require no further interaction with the terminal, e.g. double-tapping, a gesture, or otherwise. Ambient sensing has also been used in various user-device authentication, key generation and secure channel schemes [21,30]. These applications typically measure the environment for longer periods of time (>1 s) and, generally speaking, their primary goal is not proximity detection of a device with a terminal. As such, we omit these from the discussion.

2.3   Approaches and Evaluation Metrics

In previous work, two approaches have been used predominately for sensing-based PRAD mechanisms:

– *Threshold-based Similarity*: the use of time and frequency domain similarity metrics, such as Mean Absolute Error (MAE), Pearson's Correlation Coefficient and Coherence. A single threshold is generated that aims to separate all legitimate transactions from illegitimate ones using a particular similarity metric. The transaction is accepted if the metric result falls within this pre-set threshold of the maximum allowed dissimilarity.
– *Machine Learning*: the use of well-known classification algorithms, such as Naïve Bayes, Support Vector Machines (SVMs) and Random Forests. The classifier is trained on a set of feature vectors with corresponding binary labels (legitimate or relayed transaction), which are collected beforehand. The trained model is used to classify subsequent transaction data streams as legitimate or relayed.

Standard binary classification evaluation metrics have been applied to measure the effectiveness of a particular scheme, namely classification accuracy [16], f-scores[1] [32,34] and Equal Error Rate (EER) [25]. F-scores and EERs involve the computation of *false positives/acceptances* (the number of relayed transactions accepted erroneously) and *false negatives/rejections* (the number of legal transactions rejected). F-scores account for *precision*, the correct positive results divided by the number of all positive results, and *recall*, the number of correct positive results as proportion of the number of positive results that should have been identified (see Eq. 1). The EER – used extensively in biometrics, e.g. fingerprint recognition [24] – is found by calculating the False Acceptance Rate (FAR) and False Rejection Rate (FRR), shown in Eq. 2, over a range of thresholds and finding the rate at which $FAR = FRR$. Alternatively, some authors have opted to present the FAR and FRR results alone [37]. Finally, accuracy represents the correct identification of positive and negative transactions in the test set (Eq. 3), but does not clearly illustrate the number of false positives and negatives.

F-scores and accuracy have been used to primarily evaluate machine learning-based relay attack detection, e.g. [32,34], while EERs have been employed for threshold-based similarity approaches [25] to find an acceptance threshold that, broadly speaking, balances usability (false rejection rate) with security (false acceptance rate). We use the EER as a common evaluation metric for assessing the performance of machine learning and threshold-based approaches across a variety of similarity metrics.

$$F_{score} = \frac{2TP}{2TP + FP + FN} \tag{1}$$

$$FAR = \frac{FP}{FP + TN} \qquad FRR = \frac{FN}{FN + TP} \tag{2}$$

$$Accuracy = \frac{TP + TN}{P + N} \tag{3}$$

2.4 Effectiveness for Proximity Detection

In our first study [31], we evaluated the effectiveness of ambient sensors to determine whether two devices are in proximity to one another (irrespective of whether a relay attack is in action). A field trial was conducted in which sensor data from 1000 transactions per sensor was collected from 252 users at four different locations on a university campus. Two devices were used for the data collection: a transaction terminal (TT), and a transaction instrument (TI). Data was collected for 500 ms upon the initiation of the NFC-based transaction, and stored locally for later evaluation (Fig. 3).

We subjected this data to the two analyses discussed in Sect. 2.3 – threshold-based similarity and machine learning – to determine whether data from legitimately co-located devices can be distinguished from non-proximate pairs. The implementation of the test-bed, data analysis and collected data sets are made available at: https://github.com/AmbientSensorsEvaluation/ Ambient-Sensors-Proximity-Evaluation.git. The source code for the additional

---

[1] Also known as the F1 score or F-measure.

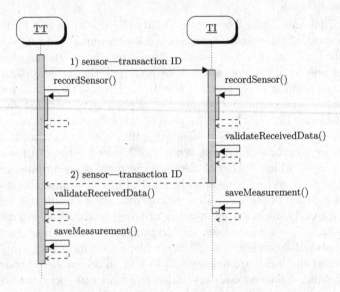

Fig. 3. Measurement recording overview.

threshold-based experiments presented in this paper can be found at: https://github.com/AmbientSensorsEvaluation/Threshold-Based-Analysis.

Analysis Approach. For each sensor, the EER and associated threshold, t, were computed using six time- and frequency-domain similarity measures, including those used in previous work. We list these forthwith. *Time domain metrics*: Mean Absolute Error (MAE), Eq. 4; Pearson's correlation coefficient [25], Eq. 5; maximum cross-correlation [16,34], Eq. 6; and Euclidean distance [34], Eq. 7. *Frequency domain*: coherence [25], Eq. 8. *Both domains*: time-frequency distance [16,34], Eq. 9. In [31], we presented results only from the MAE and Pearson's correlation coefficient; in this work, we present the results from all of these similarity metrics. Each metric was applied directly onto the sensor data collected during the field trials. For machine learning, the Weka package was employed, while a Python application was developed for threshold-based similarity learning using the Numpy, Scipy, Matplotlib and Pandas Python packages for metric implementations, graph plotting and CSV data processing.

$$MAE(A, B) = \frac{1}{N} \sum_{i=1}^{N} |A_i - B_i| \tag{4}$$

$$corr(A, B) = \frac{\sum_{i=1}^{N}((A_i - \mu_A)(B_i - \mu_B))}{\sqrt{\sum_{i=1}^{N}(A_i - \mu_A)^2 \sum_{i=1}^{N}(B_i - \mu_B)^2}} \tag{5}$$

Where μ_A represents the arithmetic mean of A.

$$M_{corr}(A, B) = \max(cross\_correlation(A, B)) \tag{6}$$

$$d(A, B) = \sqrt{\sum_{i=1}^{N} (B_i - A_i)^2} \tag{7}$$

$$C_{AB}(f) = \frac{|G_{AB}(f)|^2}{G_{AA}(f) \cdot G_{BB}(f)} \qquad F_{AB} = \sum_f C_{AB}(f) \tag{8}$$

Where G_{AA} is the auto-spectral density of A and G_{AB} is the cross-spectral density of signals A and B (left). The similarity is found by the sum of the magnitudes of coherence values at all frequencies (right).

$$Diff(A, B) = \sqrt{D_{time}(A, B)^2 + D_{freq}(A, B)^2} \tag{9}$$

Where $D_{time}(A, B) = 1 - M_{corr}(A, B)$ and $D_{freq}(A, B) = \|FFT(A) - FFT(B)\|$, in which $\|FFT(A) - FFT(B)\|$ is the Euclidean norm of the FFTs of signals A and B.

Results. The results for the threshold-based and machine learning analyses are presented in Tables 2 and 3 respectively. Note that the proximity sensor was excluded from the analysis. On some Android devices, proximity sensors return the precise distance at which an object is located from the sensor, whereas others return a binary value for whether an object is close to/far from the sensor (within 5 cm)[2]. Our test-bed devices returned only binary values. Virtually every transaction contained 'far' values, as the devices were tapped back-to-back and the sensor was located on the front of the device. Consequently, this returned identical values in almost all cases when applying the similarity metrics described previously, e.g. $MAE = 0$, which impeded threshold-finding. Machine learning was able to capture the rare times in which the sensor returned 'close' values, like when the user covered the device with their hand during the transaction. While the machine learning results are included, the issues identified mean they should be treated with caution. Other technical challenges existed elsewhere; the Rotation Vector sensor, for example, returned significant numbers of zero values on the test-bed devices, which likely distorted the results of our analysis, while sound was capable of capturing values for only half of the permitted 500 ms time-frame. The reader is referred to [31] for a breakdown of sensor success and any technical limitations encountered.

The results indicate that no sensor in either analysis can satisfactorily distinguish between proximate and non-proximate device data pairs. Some sensors provide virtually no discrimination and perform similarly to a random classifier, e.g. accelerometer (43.4–49.8% EER) and linear acceleration (42.6–50.0%). Other sensors provide better discrimination, e.g. magnetic field (29.2–32.3%) and pressure (9.2–27.0%), but still fall short of acceptable performance. Even in the best case – the pressure sensor using the Decision Tree classifier – the

---

[2] http://developer.android.com/guide/topics/sensors/sensors_position.html# sensors-pos-prox.

Table 2. Threshold-based EERs for each sensor with Mean Absolute Error (MAE), Pearson's Correlation Coefficient (PCC), Maximum Cross-Correlation (C-Corr), Euclidean Distance (ED), Coherence (Coh) and Time-Frequency Distance (T-FD). Best result for each sensor shown in bold.

Sensor	MAE	PCC	C-Corr	ED	Coh	T-FD
Accelerometer	**0.434**	0.458	0.501	0.498	0.542	0.501
GRV[a]	**0.384**	0.486	0.500	0.442	0.524	0.498
Gravity	0.429	**0.424**	0.498	0.501	0.506	0.498
Gyroscope	0.443	**0.441**	0.493	0.498	0.548	0.499
Light	0.488	0.496	0.545	0.502	**0.471**	0.546
Linear acceleration	0.496	**0.426**	0.494	0.507	0.507	0.500
Magnetic field	**0.323**	0.384	0.537	0.337	0.568	0.536
Pressure	**0.270**	0.492	0.601	0.283	0.503	0.601
Rotation Vector	0.498	0.466	0.501	0.278	0.500	**0.273**
Sound	0.417	0.488	0.481	**0.338**	0.518	0.481

Proximity excluded due to insufficient unique values.

[a] GRV: Geomagnetic Rotation Vector sensor

Table 3. Estimated EERs for machine learning algorithms, obtained by repeating stratified 10-fold cross-validation 10 times. Best result for each sensor shown in bold.

Sensor	Random Forest	Naive Bayes	Logistic Regression	Decision Tree	Support Vector Machine	Multilayer Perceptron
Accelerometer	0.626 ± 0.024	0.509±0.026	0.526±0.023	0.500 ± 0.0	**0.498** ± 0.025	0.551± 0.025
GRV	**0.435**± 0.021	0.447 ± 0.024	0.474 ± 0.031	0.500 ± 0.0	0.489 ± 0.036	0.450 ± 0.026
Gravity	0.874 ± 0.018	0.579 ± 0.020	0.579 ± 0.024	**0.500** ± 0.0	**0.500** ± 0.026	0.746 ± 0.112
Gyroscope	0.683 ± 0.027	**0.499** ± 0.024	0.543 ± 0.024	0.500 ± 0.0	0.511 ± 0.025	0.514 ± 0.025
Light	0.576 ± 0.026	0.515 ± 0.024	0.533 ± 0.025	**0.500** ± 0.0	0.508 ± 0.024	0.513 ± 0.028
Linear acceleration	0.603 ± 0.025	0.507 ± 0.027	0.543 ± 0.023	**0.500** ± 0.0	**0.500** ± 0.021	0.554 ± 0.028
Magnetic field	**0.292** ± 0.021	0.319 ± 0.020	0.322 ± 0.020	0.415 ± 0.015	0.398 ± 0.046	0.329 ± 0.026
Pressure	0.103 ± 0.010	0.107 ± 0.010	0.287 ± 0.013	**0.092** ± 0.054	0.319 ± 0.045	0.114 ± 0.019
Proximity	0.499 ± 0.031	0.537 ± 0.069	**0.476** ± 0.188	0.500 ± 0.0	0.543 ± 0.254	0.508 ± 0.197
Rotation Vector	**0.276** ± 0.046	0.563 ± 0.243	0.596 ± 0.233	0.500 ± 0.0	0.513 ± 0.243	0.488 ± 0.245
Sound	**0.288** ± 0.019	0.314 ± 0.022	0.310 ± 0.021	0.347 ± 0.136	0.411 ± 0.041	0.306 ± 0.020

EER was 9.2%. By definition of the EER, this implies that approximately 9.2% of both legitimate and illegitimate transactions would be rejected and accepted respectively. Rejecting almost 1-in-10 legitimate transactions in a high throughput scenario is likely to cause user annoyance in practice, such as mobile ticketing in a subway system. As such, it is difficult to recommend any single sensor in our analysis as an effective proximity detection method.

2.5 Effectiveness for Relay Attack Detection

The first evaluation [31] focused only on proximity detection, rather than using data from relay attacks. In our next major work, we conducted further field trials [14] in which data was collected from two devices that were genuinely in proximity, and a third device that was located 1.5 m/5 ft away. This replicated a relay attack in which an adversary launches the attack on a nearby victim, such as in a shop queue. We aimed to determine whether sensor data from the relay device pair – the terminal and the device 5 ft away – could be distinguished from a legitimate pair, i.e. the terminal and the device in proximity.

The relay pair comprised a transaction terminal (TT) and a transaction instrument (TI′), whereas the legitimate pair consisted of a relay transaction terminal (TT′), and a transaction instrument (TI). Devices TI′ and TI were tapped simultaneously against devices TT and TT′ respectively. A 500 ms NFC-based transaction was then initiated on both sides and, upon completion, the devices TT, TI′, and TI stored the collected sensor data locally for off-line analysis. Figure 4 presents an overview of the recording process.

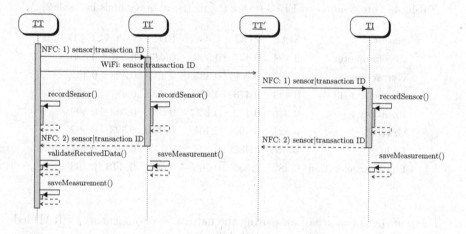

Fig. 4. Measurement recording process.

The implementation of the test-bed, data analysis and collection data sets for [14] are available at: https://github.com/AmbientSensorsEvaluation/Ambient-Sensors-Relay-Attack-Evaluation.

Analysis Approach and Results. In this study, we limited our sensor selection to the best performing sensors from our first analysis (Shepherd et al. [31]), i.e. those which successfully and consistently captured values within the 500 ms time limit over 1,000 transactions. Additionally, based on our initial investigations, we eliminated sensors that are largely uncommon on commodity handsets in the current market (2016–2017), like the pressure sensor. The reader is referred to [14,31] for a detailed discussion of these matters. The same analysis techniques were used as the previous work for proximity detection – threshold-based analysis with the same similarity measures, and machine learning – using the EER evaluation metric. The results for the threshold-based and machine learning analyses of this study can be found in Tables 4 and 5 respectively.

Similar to our previous study (Sect. 2.4), some sensors provided poor discriminatory power; the magnetic field sensor, for instance, gave EERs of 36.1% and 43.3% in the analyses. Some sensors provided greater discriminatory power, e.g. gyroscope (17.9% EER with Random Forest) and the rotation vector sensor (27.7%, also with Random Forest). These EERs, however, are still too high to recommend as an effective PRAD in high throughput situations. Based on this evaluation, we reached the tentative conclusion that sensing the transaction devices' natural ambient environment may not be a suitable PRAD mechanism under industry-specified time constraints. In future work, we aim to conduct additional experiments using multiple sensors with various permutations and sensor fusion techniques to further interrogate the veracity of our conclusions.

Table 4. Threshold-based EERs (using the metric abbreviations in Table 2).

Sensor	MAE	PCC	C-Corr	ED	Coh	T-FD
Accelerometer	0.494	0.477	0.590	**0.468**	0.507	0.590
Gyroscope	0.521	**0.455**	0.535	0.495	0.528	0.489
Magnetic field	0.444	0.473	0.470	**0.433**	0.487	0.470
Rotation vector	0.330	0.472	**0.327**	0.670	0.534	0.509
Gravity	0.521	0.490	0.401	**0.289**	0.503	0.362
Light	**0.367**	0.488	0.444	0.372	0.505	0.437
Linear acceleration	0.482	0.536	0.503	0.506	**0.443**	0.493

The poor performance of measuring the natural environment as a PRAD led us to explore the generation of an artificial ambient environment that is unique to each transaction. In the following section of this paper, we introduce and discuss artificial ambient environments in greater detail.

3   Detection via Artificial Ambient Environments

As an alternative to the natural ambient environment, we proposed the generation of an Artificial Ambient Environment (AAE) using the peripherals of the transaction devices [12]. In this section, we discuss the basic principles of using AAEs as an anti-relay mechanism. Firstly, we present how infrared light can be used as an AAE actuator, before describing the use of sound as a proximity detection mechanism and as a communication medium for proximate devices. Finally, we suggest how other actuators may be used to provide proximity assurances.

Table 5. Estimated EER for machine learning algorithms, obtained by repeating 10-fold cross-validation 10 times.

Sensor	Random Forest	Naive Bayes	Decision Tree	Logistic Regression	Support Vector Machine
Accelerometer	**0.277** ± 0.052	0.474 ± 0.047	0.358 ± 0.059	0.483 ± 0.050	0.454 ± 0.126
Gyroscope	**0.179** ± 0.041	0.354 ± 0.059	0.228 ± 0.049	0.356 ± 0.055	0.288 ± 0.045
Magnetic field	**0.361** ± 0.055	0.400 ± 0.053	0.389 ± 0.063	0.421 ± 0.061	0.385 ± 0.053
Rotation vector	**0.285** ± 0.052	0.327 ± 0.055	0.317 ± 0.073	0.353 ± 0.050	0.325 ± 0.050
Gravity	0.499 ± 0.046	0.488 ± 0.043	0.494 ± 0.057	**0.484** ± 0.043	0.486 ± 0.156
Light	0.361 ± 0.059	0.369 ± 0.058	**0.293** ± 0.149	0.407 ± 0.054	0.351 ± 0.054
Linear acceleration	**0.307** ± 0.050	0.484 ± 0.048	0.392 ± 0.057	0.502 ± 0.049	0.397 ± 0.058

3.1   Artificial Ambient Environments

In order to increase the irreproducibility and uniqueness of an ambient environment, the transaction devices generate an artificial environment using their peripherals – measurable by a particular ambient sensor(s). The artificial environment should be based on randomly generated bits or sequences to act as a second (out-of-band) channel for assuring proximity between the transaction devices (see Fig. 5).

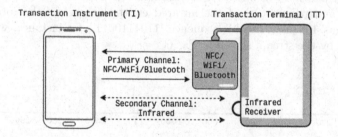

Fig. 5. High-level communication overview.

Upon initiation of a transaction, one (unidirectional) or both (bidirectional) device(s) are responsible for the generation and/or sensing of the AAE for some predefined time. After recording the sensor measurement, a comparison is performed with the captured data from both devices. The comparison may be performed by one of the communicating parties or by a trusted third party, as discussed in Sect. 2.1.

During the comparison, only the data that was captured while the artificial ambient channel was active should be considered; data captured outside this time-frame should be discarded. This way, an attacker cannot capture the generated sequence and then replay it at a remote location. Moreover, for an effective

AAE, the attacker should not be able to relay the data from the out-of-band channel in a way that the trusted comparison party cannot distinguish between a legitimate and an illegitimate transaction with a high degree of confidence.

To summarise, the basic principles of an AAE are:

1. The AAE generation should be based on random bits/sequences.
2. The AAE should provide sufficient evidence in order for two genuine devices to establish proximity assurance.
3. It should be hard for the attacker to accurately reproduce the AAE at a remote location.

The primary goal of the AAE is to protect against the off-the-shelf attacker. A resourceful attacker with access to state of the art equipment might be capable of effectively reproducing the same conditions at a remote location in a timely fashion. However, smartphones suffer from a plethora of security issues [26] and, in practice, a resourceful attacker is more likely to exploit these than invest in state of the art hardware to conduct a relay attack. On modern smartphones, widely-available peripherals that could potentially act as AAE actuators include: 1. the device's infrared emitter, 2. speaker, 3. flash light, 4. vibration, 5. display, 6. WiFi, 7. Bluetooth, 8. camera.

3.2   Infrared Light as an AAE Actuator

In [12], the use of infrared light as an AAE actuator was empirically evaluated. The AAE generation was based on 500 random bits, represented by $200\,\mu s$ long pulses (1s) and pauses (0s) of the infrared emitter (therefore the total emission time was 100 ms). The bit sequence '1101110011', for instance, would be represented by the stream shown in Fig. 6.

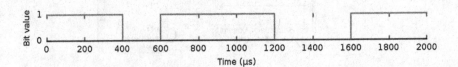

Fig. 6. Representation of the bit sequence '1101110011' in pulses-pauses.

The transaction instrument (TI) begins the infrared emission process when the transaction is initiated. The transaction terminal (TT) listens for infrared signals for 100 ms plus some acceptable offset window (4 ms), and rejects any signals received outside of this time-frame. Due to intrinsic hardware delays encountered during the experiments, TI was not able to immediately initiate the emission process, and some time x_i was required prior to the process to compensate for this. This time was quantifiable because the total emission time (100 ms) was known, as well as the total time required between the initiation and completion of the emission process. The bits accepted by device TT hence depended upon time x_i. Any bits captured prior to $(x_i - 2$ ms$)$ and after $(x_i + 100$ ms $+ 2$ ms$)$ were rejected, where the 2 ms before and after comprises the acceptable offset. The offset is the maximum allowed deviation from the average time required by the transaction initiation. Figure 7 depicts the process on the two channels.

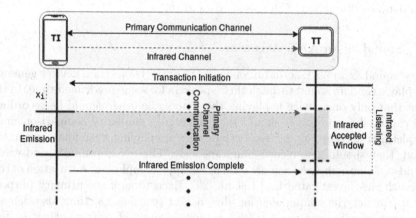

Fig. 7. Infrared as an AAE actuator.

The generated bits from TI and the captured bits from TT are compared for similarity after this process. As mentioned in Sect. 2.1, the comparison itself may take place on the terminal or by a trusted third party. Further investigations [13] showed that the process overhead is minimal, and the comparison can take place effectively during the transaction time (500 ms) by one of the devices.

The assumption of this technique is that an attacker, using off-the-shelf equipment, cannot effectively relay infrared data within that time-frame without being detected. A delay of more than 200 μs in relaying a single bit would introduce new bits into the sequence. Caching might increase the potential for an attacker to evade the proposed solution, since the risk of introducing extra bits is shifted from the bit level to the cache length level (when the relay is delayed by more than 200 μs). However, extensive caching (more than approximately 4 ms) would prolong the completion of the emission process beyond the acceptable time-limit of 100 ms, and so the delayed bits would not be considered. Hence, caching and subsequently relaying segments of the captured random sequence would have to be limited a maximum of a few milliseconds. Moreover, relay equipment such as fibre optics, where a cable connects the two relay devices, were not considered, as it can be easily detectable by the victim and/or terminal operator.

During a legitimate transaction, the similarity between the emitted and captured data was measured to be 98% or higher on approximately 98% of the performed transactions. This, therefore, was set as a baseline threshold. The dwdiff tool[3] was used for the comparison of the two bit-streams. Six distinct relay test-beds were developed using off-the-shelf equipment, such as infrared extenders, Raspberry Pis, and mobile devices. None of the test-beds could effectively attack the proposed solution. The highest similarity rate after conducting a relay attack was achieved by an infrared extender, with 95% similarity across approximately 10% of all performed transactions. The reader is referred to [12] for a detailed discussion of the test-beds and the results.

3.3 Sound as an AAE Actuator

To use sound as an AAE actuator, one or both of the transaction devices generate and play a random sound through their speakers for some predefined time. In the event that only one device is playing the sound, the other should be recording. The captured sound-waves should then undergo a similarity comparison upon completing the recording. In case both devices are playing a randomly generated sound, they should also record simultaneously. The two captured sound-waves should again be compared against each other upon completion. A variation of this approach was investigated by Li et al. [22]. Even though the primary purpose was to restrict the communicating distance of two devices, they also demonstrate the effectiveness against relay attacks as part of their solution. In this work, the acoustic channel was used as the main channel for communicating messages between two devices. Signals were transmitted at a high frequency by both devices simultaneously (full-duplex communication). The full-duplex approach assisted towards relay attack prevention, as each device was capturing signals from both communicating devices, including itself. It could therefore estimate whether a message from the opposite party was received within some acceptable window by juxtaposing it with the message emanating from itself. Positive results were reported by the authors with a high success rate in silent environments; however, this degraded as the surrounding environment became noisier.

Karapanos et al. [21] explored the use of sound as a two-factor authentication method; however, attacks and potential solutions against this method have been demonstrated by Shrestha et al. [33].

Yi et al. [39] used the acoustic channel as a means of user authentication for unlocking mobile devices using a wearable device (smartwatch). The authors reported promising results with a lower bit error rate than conventional smartphone unlocking methods, e.g. PIN entry. While the main purpose of this work was not to counter relay attacks, they argue that such attacks would be expensive to carry out and, as the size of the relay equipment is large, it could easily be identified by a genuine user. They also claim that fingerprinting techniques can be used to uniquely identify the acoustic hardware to determine whether it originated from a genuine or relay device. Lastly, they mention that distance bounding protocols can be used, but a full investigation of this was considered out of scope.

---

[3] dwdiff tool: http://os.ghalkes.nl/dwdiff.html.

Based on the analysis in [31], while promising results have been demonstrated through the use of sound as a PRAD mechanism, it might not be applicable in EMV, transport ticketing, or other transactions with industry-imposed time restrictions of up to 500 ms. On average, due to latency related to initiating the recording process, recordings lasted for less than 280 ms within a 500 ms permitted time-frame. Similar latencies were not observed on most of the evaluated sensors; we concluded that sensor hardware may have a significant bearing on the effectiveness of a PRAD mechanism.

3.4   Other AAE Actuators

In this section, we discuss other potential AAE actuators; we focus on the candidate actuators listed in Sect. 3.1. In some cases, like in the case the Bluetooth or the WiFi, the underlying technology may not be flexible enough to be used as an AAE actuator without substantial modifications.

The display of a device could be used as a potential AAE actuator in combination with the camera of the communicating device. One device could display a randomly generated video feed to be recorded by the other device; the displayed and captured video feeds may then be compared for similarity. The advantage of this technique is that relaying video may incur a relatively large degree of latency. The main downsides are that: 1. the two devices ought to be held in the correct way to maximise success; and 2. the delay in capturing and playback of a stream could potentially negatively affect the results.

Similarly, the device's flashlight could be used by displaying a random pattern that is captured by the communicating party's camera or light sensor. Previous work has achieved reliable emission at speeds up to 500 bits per second (bps) using a LED flashlight and 15 bps using a Xenon flashlight [10,11,19], but this is 5 times slower than a typical infrared emitter, as per [12]. As such, an attacker would have a much larger relay window, which may hinder the effectiveness of this method. Additionally, the flashlight falls within the visible light spectrum, which may physically disturb nearby users.

One other potential option is vibration. While vibration has not been used to the best of our knowledge as a relay attack detection mechanism, it has been used to authenticate RFID tags [28] and to exchange secrets [29] with a high success rate. This evidence suggests that there is a potential in using vibration as an anti-relay mechanism. Here, one or both of the transaction devices generate a random vibration pattern, which is measured by both devices using a motion sensor, e.g. accelerometer. The captured data can be transmitted to a trusted third party upon the completion of the transaction for comparison.

4   Conclusion and Future Work

Proximity and Relay-Attack Detection (PRAD) is an important element for many contactless and wireless technologies. In this work, we illustrated that the viability of PRAD mechanisms can be largely dependent on the time constraints mandated by industry requirements. Contactless payment transactions for example – whether smart card- or smartphone-based – must adhere to <3 cm for proximity and <500 ms for transaction duration, as stipulated by the EMV specifications. We evaluated the claim that natural ambient environments can provide a robust PRAD, as stated by some previous literature, under industry-specified time constraints. This was evaluated for both proximity detection (Sect. 2.4) and as a relay attack detection mechanism using a test-bed that reflected an actual attack (Sect. 2.5). We presented the results of a two-part evaluation using six similarity metrics used previously and several widely-used machine learning classifiers. In all cases, the results were far from what was claimed in past literature; our initial results indicate that natural ambient environments provide a poor PRAD for time-critical domains such as banking and transport. As such, we strongly recommend that any PRAD proposal should be evaluated based on the operating restrictions of the suggested deployment application.

This led to the development of artificially generated environments, which are random and unique to each transaction, for providing a more effective PRAD mechanism. To test this, we proposed a framework for deploying an artificial ambient environment (AAE) for PRAD. We developed a test-bed to evaluate the effectiveness of infrared in conjunction with six relay attack test-beds. In all cases, the genuine and relayed transactions were distinguishable for 97–98% of all transactions – far greater than the results using natural ambient sensing from our investigations. At present, we are expanding our interrogation of natural environment-based PRADs, using multiple sensors simultaneously with a range of sensor fusion techniques. Moreover, we are investigating the applicability of other smartphone sensors as an AAE-based PRAD mechanism. The first phase of this evaluation has been conducted using vibration as an AAE, which has yielded promising results.

Acknowledgement. Carlton Shepherd is supported by the EPSRC and the British government as part of the Centre for Doctoral Training in Cyber Security at Royal Holloway, University of London (EP/K035584/1). The authors would also like to thank anonymous reviewers for their valuable comments.

References

1. Transit and Contactless Open Payments: An Emerging Approach for Fare Collection. White paper, Smart Card Alliance Transportation Council, November 2011
2. How to Optimize the Consumer Contactless Experience? The Perfect Tap. Technical report, MasterCard (2014)
3. EMV Contactless Specifications for Payment Systems: Book D - EMV Contactless Communication Protocol Specification. Spec V2.6, EMVCo, LLC, March 2016
4. Transactions Acceptance Device Guide (TADG). Specification Version 3.1, VISA, November 2016
5. Boureanu, I., Mitrokotsa, A., Vaudenay, S.: Towards secure distance bounding. In: Moriai, S. (ed.) FSE 2013. LNCS, vol. 8424, pp. 55–67. Springer, Heidelberg (2014). doi:10.1007/978-3-662-43933-3_4

6. Coskun, V., Ozdenizci, B., Ok, K.: A survey on Near Field Communication (NFC) technology. Wireless Pers. Commun. **71**(3), 2259–2294 (2013). http://dx.doi.org/10.1007/s11277-012-0935-5
7. Cremers, C., Rasmussen, K., Schmidt, B., Capkun, S.: Distance hijacking attacks on distance bounding protocols. In: 2012 IEEE Symposium on Security and Privacy, pp. 113–127, May 2012
8. Francis, L., Hancke, G., Mayes, K., Markantonakis, K.: Practical NFC peer-to-peer relay attack using mobile phones. In: Ors Yalcin, S.B. (ed.) RFIDSec 2010. LNCS, vol. 6370, pp. 35–49. Springer, Heidelberg (2010). doi:10.1007/978-3-642-16822-2_4
9. Francis, L., Hancke, G.P., Mayes, K., Markantonakis, K.: Practical relay attack on contactless transactions by using NFC mobile phones. In: IACR Cryptology Archive 2011, p. 618 (2011)
10. Galal, M.M., Fayed, H.A., Aziz, A.A.E., Aly, M.H.: Smartphones for payments and withdrawals utilizing embedded LED flashlight for high speed data transmission. In: 2013 Fifth International Conference on Computational Intelligence, Communication Systems and Networks, pp. 63–66, June 2013
11. Galal, M.M., Aziz, A.A.A.E., Fayed, H.A., Aly, M.H.: Smartphone payment via flashlight: utilizing the built-in flashlight of smartphones as replacement for magnetic cards. Optik - Int. J. Light Electron Optics **127**(5), 2453–2460 (2016)
12. Gurulian, I., Akram, R.N., Markantonakis, K., Mayes, K.: Preventing relay attacks in mobile transactions using infrared light. In: Proceedings of the Symposium on Applied Computing, SAC 2017, pp. 1724–1731. ACM, New York (2017)
13. Gurulian, I., Markantonakis, K., Akram, R.N., Mayes, K.: Artificial ambient environments for proximity critical applications. In: 2017 12th International Conference on Availability, Reliability and Security, ARES 2017. ACM, New York (2017)
14. Gurulian, I., Shepherd, C., Frank, E., Markantonakis, K., Akram, R., Mayes, K.: On the effectiveness of ambient sensing for NFC-based proximity detection by applying relay attack data. In: The 16th IEEE International Conference on Trust, Security and Privacy in Computing and Communications, TrustCom 2017. IEEE, August 2017
15. Haken, G., Markantonakis, K., Gurulian, I., Shepherd, C., Akram, R.N.: Evaluation of Apple iDevice sensors as a potential relay attack countermeasure for Apple Pay. In: Proceedings of the 3rd ACM Workshop on Cyber-Physical System Security, CPSS 2017, pp. 21–32. ACM, New York (2017)
16. Halevi, T., Ma, D., Saxena, N., Xiang, T.: Secure proximity detection for NFC devices based on ambient sensor data. In: Foresti, S., Yung, M., Martinelli, F. (eds.) ESORICS 2012. LNCS, vol. 7459, pp. 379–396. Springer, Heidelberg (2012). doi:10.1007/978-3-642-33167-1_22
17. Hancke, G.P., Kuhn, M.G.: Attacks on time-of-flight distance bounding channels. In: Proceedings of the First ACM Conference on Wireless Network Security, WiSec 2008, pp. 194–202. ACM, New York (2008). http://doi.acm.org/10.1145/1352533.1352566
18. Hancke, G., Mayes, K., Markantonakis, K.: Confidence in smart token proximity: relay attacks revisited. Comput. Secur. **28**(7), 615–627 (2009). http://www.sciencedirect.com/science/article/pii/S0167404809000595
19. Hesselmann, T., Henze, N., Boll, S.: FlashLight: optical communication between mobile phones and interactive tabletops. In: ACM International Conference on Interactive Tabletops and Surfaces, ITS 2010, pp. 135–138. ACM, New York (2010), http://doi.acm.org/10.1145/1936652.1936679

20. Jin, R., Shi, L., Zeng, K., Pande, A., Mohapatra, P.: MagPairing: pairing smartphones in close proximity using magnetometers. IEEE Trans. Inf. Forensics Secur. **11**(6), 1306–1320 (2016)
21. Karapanos, N., Marforio, C., Soriente, C., Capkun, S.: Sound-Proof: usable two-factor authentication based on ambient sound. In: 24th USENIX Security Symposium. USENIX Association, Washington, D.C., August 2015
22. Li, L., Xue, G., Zhao, X.: The power of whispering: near field assertions via acoustic communications. In: Proceedings of the 10th ACM Symposium on Information, Computer and Communications Security, ASIA CCS 2015, pp. 627–632. ACM, New York (2015). http://doi.acm.org/10.1145/2714576.2714586
23. Ma, D., Saxena, N., Xiang, T., Zhu, Y.: Location-aware and safer cards: enhancing RFID security and privacy via location sensing. IEEE TDSC **10**(2), 57–69 (2013)
24. Maltoni, D., Maio, D., Jain, A., Prabhakar, S.: Handbook of Fingerprint Recognition. Springer Science & Business Media, London (2009). doi:10.1007/978-1-84882-254-2
25. Mehrnezhad, M., Hao, F., Shahandashti, S.F.: Tap-Tap and Pay (TTP): preventing man-in-the-middle attacks in NFC payment using mobile sensors. In: 2nd International Conference on Research in Security Standardisation, October 2014
26. Polla, M.L., Martinelli, F., Sgandurra, D.: A survey on security for mobile devices. IEEE Commun. Surv. Tutorials **15**(1), 446–471 (2013)
27. Rasmussen, K.B., Capkun, S.: Realization of RF distance bounding. In: USENIX Security Symposium, pp. 389–402 (2010)
28. Saxena, N., Uddin, M.B., Voris, J., Asokan, N.: Vibrate-to-unlock: mobile phone assisted user authentication to multiple personal RFID tags. In: 2011 IEEE International Conference on Pervasive Computing and Communications (PerCom), pp. 181–188, March 2011
29. Shen, Z., Zheng, X., Xie, H.: Near field service initiation via vibration channel. In: 2016 12th International Conference on Mobile Ad-Hoc and Sensor Networks (MSN), pp. 450–453, December 2016
30. Shepherd, C., Akram, R.N., Markantonakis, K.: Towards trusted execution of multi-modal continuous authentication schemes. In: Proceedings of the 32nd Symposium on Applied Computing, pp. 1444–1451. ACM (2017)
31. Shepherd, C., Gurulian, I., Frank, E., Markantonakis, K., Akram, R., Mayes, K., Panaousis, E.: The applicability of ambient sensors as proximity evidence for NFC transactions. In: Mobile Security Technologies, IEEE Security and Privacy Workshops, MoST 2017. IEEE, May 2017
32. Shrestha, B., Saxena, N., Truong, H.T.T., Asokan, N.: Drone to the rescue: relay-resilient authentication using ambient multi-sensing. In: Christin, N., Safavi-Naini, R. (eds.) FC 2014. LNCS, vol. 8437, pp. 349–364. Springer, Heidelberg (2014). doi:10.1007/978-3-662-45472-5_23
33. Shrestha, B., Shirvanian, M., Shrestha, P., Saxena, N.: The sounds of the phones: dangers of zero-effort second factor login based on ambient audio. In: Proceedings of the 2016 ACM SIGSAC Conference on Computer and Communications Security, CCS 2016 pp. 908–919. ACM, New York (2016)
34. Truong, H.T.T., Gao, X., Shrestha, B., Saxena, N., Asokan, N., Nurmi, P.: Comparing and fusing different sensor modalities for relay attack resistance in zero-interaction authentication. In: 2014 IEEE International Conference on Pervasive Computing and Communications, pp. 163–171. IEEE (2014)
35. Umar, A., Mayes, K., Markantonakis, K.: Performance variation in host-based card emulation compared to a hardware security element. In: 2015 First Conference on Mobile and Secure Services, pp. 1–6. IEEE (2015)

36. Urien, P., Piramuthu, S.: Elliptic curve-based RFID/NFC authentication with temperature sensor input for relay attacks. Decision Support Syst. **59**, 28–36 (2014)
37. Varshavsky, A., Scannell, A., LaMarca, A., de Lara, E.: Amigo: proximity-based authentication of mobile devices. In: Krumm, J., Abowd, G.D., Seneviratne, A., Strang, T. (eds.) UbiComp 2007. LNCS, vol. 4717, pp. 253–270. Springer, Heidelberg (2007). doi:10.1007/978-3-540-74853-3_15
38. Verdult, R., Kooman, F.: Practical attacks on NFC enabled cell phones. In: 2011 3rd International Workshop on Near Field Communication (NFC), pp. 77–82, February 2011
39. Yi, S., Qin, Z., Carter, N., Li, Q.: WearLock: unlocking your phone via acoustics using smartwatch. In: 2017 IEEE 37th IEEE International Conference on Distributed Computing Systems, ICDCS 2017 (2017)

Challenges of Federating National Data Access Infrastructures

Margus Freudenthal and Jan Willemson[✉]

Cybernetica, Ülikooli 2, Tartu, Estonia
{margus,janwil}@cyber.ee

Abstract. X-Road is a secure and scalable database access middleware originally developed in Estonia in early 2000s. In 2014, a decision was taken to also deploy X-Road infrastructure within Finland, hence facilitation cross-national federation. Even though being very close both geographically and culturally, the legislation, technology and best practices used by the two nations differ. This paper discusses the nature and implications of these differences in the context of federated installation of the infrastructure.

Keywords: Secure database access · Cross-national security infrastructure federation

1 Introduction

By late 1990s, the level of computerization in both public and private sectors had reached the stage where large-volume digital data exchange between organizations became both feasible and necessary. Various government registries implemented electronic interfaces that could be used to query data from the registry. However, these interfaces suffered from two problems.

First, each registry implemented the interfaces independently, often using a proprietary protocol implied by the technology used. Hence, when an organization needed interfaces to several other registries, new interfaces had to be implemented for each one of them almost from scratch.

Second, as digital data exchange became more widespread, security aspects of the queries required more and more attention. On one hand, registries often contain confidential personal data, hence access to it must be tightly controlled. On the other hand, the client requests the data to make a (possibly costly) decision based on it. Hence it must be possible to verify the integrity and authenticity of the received data. However, the registry interfaces had varying levels of security, depending on the implementer.

Thus there was clear need to enhance standardization both from interoperability as well as security point of view. To achieve this goal in Estonia, a unified data exchange middleware called X-Road was launched in December 2001 [7,9]. During the next years it has evolved with addition of features and evolutionary changes to protocols and data formats, reaching version 5 in 2010.

© Springer International Publishing AG 2017
P. Farshim and E. Simion (Eds.): SecITC 2017, LNCS 10543, pp. 104–114, 2017.
https://doi.org/10.1007/978-3-319-69284-5_8

By the end of 2016, X-Road had 1789 connected services by 246 service providers. Altogether, 975 member organizations exchanged roughly 575 million transactions per year[1]. For comparison – the population of Estonia is slightly over 1.3 million which gives more than 430 transactions per inhabitant per year.

Development of next generation of X-Road started in 2014. It was based on product prototype developed two years earlier by Cybernetica AS, the developer and maintainer of the original X-Road software. One of the goals for new version was better support for international deployment and cross-border electronic services. The new version, version 6, was a fresh start and did not use the same technical solution and protocols as the previous versions 1 to 5.

X-Road version 6 was also licensed to Finnish government and the source code was published on Github[2]. Currently it is maintained in cooperation by Estonian and Finnish governments. X-Road version 6 is being implemented in both Estonia and Finland.

There is also a commercial branch of X-Road developed by Cybernetica, the company responsible for development and maintenance of the previous X-Road versions. It is called Cybernetica UXP® (Unified eXchange Platform) and is based on the same product prototype as X-Road version 6. UXP includes improved versions of X-Road components as well as additional components that simplify implementation in other countries. As of 2017, this product has been installed in Haiti, Namibia and also in a pilot environment in the United Kingdom. UXP uses the same data formats and message standards as X-Road, maintaining full service level compatibility.

Since X-Road technology is already deployed in several countries, one major prerequisite for implementing cross-border digital services is fulfilled. However, actual implementation of such services introduces an extra layer of complexity even for countries that are culturally close and otherwise friendly (such as as Estonia and Finland). For example, there are still noticeable differences in the legal systems and trust levels provided by the trust service providers (like certification authorities) may be incompatible.

The aim of the current paper is to describe technical implementation of X-Road federation and explore the problems that may arise as a result of the differences between federated installations.

2   X-Road Infrastructure

General structure of X-Road infrastructure as deployed in one country is described in detail in [8]. In this paper, we will provide a concise overview; see Fig. 1.

An X-Road installation is managed by a *Governing Authority*. This is the body responsible for determining the legal status as well as overall policies concerning the data exchange. The Governing Authority manages the members of the X-Road installation. In case of federation, Governing Authorities will also

---

[1] https://www.ria.ee/ee/x-tee-statistika.html.
[2] https://github.com/ria-ee/X-Road.

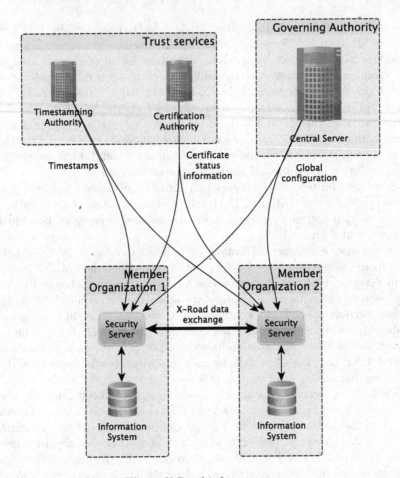

Fig. 1. X-Road infrastructure

serve as national contact points, establishing bilateral agreements and everything else needed to ensure interoperability.

The Governing Authority is responsible for setting up and maintaining a *Central Server*. The Central Server contains a member directory and other data. This data is distributed to organizations as global configuration (see Sect. 2.1 for a more detailed discussion).

All the messages exchanged over X-Road are digitally signed to provide both integrity and non-repudiation properties. Hence, X-Road assumes a Public Key Infrastructure (PKI) to function properly. The main PKI components required by the system are Certification Authority and Timestamping Authority. The Governing Authority specifies a list of the trusted PKI service providers and this list is distributed as part of global configuration.

In X-Road interoperability layer, the *X-Road member organizations* communicate directly with each other for data transfer. The communication is structured as synchronous service calls. The data exchange uses mutually authenticated TLS as the transport protocol. All messages carry proof value that is created by signing and timestamping all the exchanged messages.

X-Road members are very different organizations varying from small companies to large governmental institutions. Accordingly, their IT capabilities also vary. Together with different technologies used by every organization, achieving a standardized set of well-implemented security measures is very difficult. Instead of specifying the security protocols and relying on the members to correctly implement it, X-Road uses standard components called *security servers*. Security servers encapsulate the X-Road security protocol and ensure that it is implemented properly. They act as gateways between the organization's information system and the X-Road infrastructure (see Fig. 1). Security server software is developed centrally and distributed to the member organizations.

Whereas data exchange takes place directly between the member organizations' security servers, the decentralized system is governed by a central Governing Authority. Besides maintaining registry of the members and security servers, the Governing Authority defines, distributes and enforces policies for the whole system, e.g. security policies. The security policy of an X-Road instance consists of the following items:

- list of trusted certification authorities,
- list of trusted timestamping authorities,
- some tunable security parameters such as
  - maximum allowed lifetime of an OCSP response (how often the certificate validity information must be refreshed),
  - maximum allowed granularity of time-stamping confirmations (how much the time in the cryptographic timestamp can differ from message time).

Security policies are, together with some other management information (like the X-Road member directory), distributed as part of *global configuration*. Since all the trust that X-Road members have towards the whole infrastructure relies on authenticity and integrity of this configuration, its distribution is the most security-critical operation during the initialization process.

2.1   Configuration Management

Global configuration is distributed by the central server to X-Road members in a set of signed XML files. However, the corresponding verification key can not come as a part of this configuration, but must enter the system from a different, *a priori* trusted source. In case of X-Road, this trusted source is established by loading the verification keys manually to the security servers in form of *configuration anchor* files.

A configuration anchor file contains a URL that can be used to download the global configuration, and a set of public keys to verify authenticity of the downloaded files. The configuration anchor is distributed via out of band means and loaded into the security server on initialization. The anchor is then used to verify the downloaded global configuration (containing approved CA certificates) that, in turn, is used to verify certificates used by security servers.

In case of federated X-Road installations, configuration is typically split into two parts (see [6]):

- *private parameters*: set of parameters that are used only by members of this X-Road instance (for example, addresses of certain management services), and
- *shared parameters*: set of parameters that are used by members of this X-Road instance and other federated instances (for example, member directory, list of trusted CAs).

For added flexibility, private parameters can include additional configuration anchors. These anchors can either refer to shared parameters of the same instance (typically both configuration files are served from the same URL endpoint) or some other X-Road instance. This mechanism is used to set up federation relationships between X-Road instances.

Figure 2 shows an example configuration with two X-Road instances. Here the security servers are initialized with a freestanding configuration anchor. Using this anchor, they can download and verify a private parameters file. The anchors in the private parameters file can, in turn, be used to download shared parameters files of both local and remote X-Road instances. Note that the configuration anchors cannot be chained – the security servers only trust anchors found from the private parameters file of their own instance.

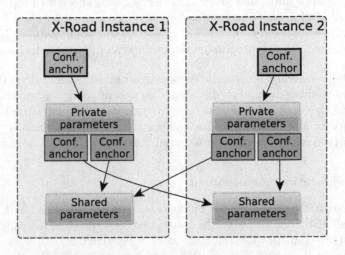

Fig. 2. X-Road configuration management

3   Implementing X-Road Federation

The options for federating different X-Road installations were first studied by Ansper and Willemson in 2008 [11]. Three possible strategies were proposed:

1. A new higher level is defined having all the present X-Road infrastructures as its descendants.
2. To facilitate international queries, a new cross-border X-Road instance is established in parallel with the existing ones.
3. All nations have their own X-Road infrastructures, and no additional ones are defined. In order to allow international information exchange, bilateral agreements are made between the existing governing institutions.

This paper explores more closely the third option that is also selected for federating Estonian and Finnish X-Road instances.

X-Road federation infrastructure is depicted in Fig. 3.

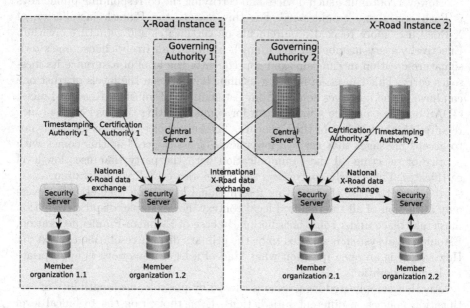

Fig. 3. X-Road federation

Here the two governing authorities enter into a bilateral agreement. Based on this, they exchange the configuration anchors pointing to their respective shared parameters file. Each governing authority copies the anchor of their partner into their private parameters file so that all the security servers can download and verify the configuration of the federated instance. X-Road uses the PKI in such a manner that each security server only interacts with the Certification Authority that issued its own certificates. Every time a security server receives a certificate, it comes with all the necessary information (OCSP responses for all the certificates in the certificate chain) needed to verify it. This method is also used in the federated setting – security servers do not make any requests to trust services of another federated instance; they only use list of trusted CA certificates from global configuration to verify certificates.

4   Legal Challenges

Reaching bilateral agreements between two governing authorities is always non-trivial. Even though huge efforts are put to trust service level unification on the European Union level, there is still a lot of room for discrepancies.

For example, the Regulation No 910/2014 of the European Parliament and of the Council of 23 July 2014 on electronic identification and trust services (known as eIDAS regulation [1]) defines several trust levels of digital certificates. In the Estonian X-Road, the signatures created by security servers comply with the requirements for a qualified electronic seal. This, among other things, means that qualified signature creation devices are required for all the X-Road members [4].

However, obtaining such devices and certifying the corresponding public keys is both costly and troublesome. For this reason, the Finnish installation of X-Road uses more relaxed requirements on certificates and signature creation. Effectively, every member of the Finnish instance can freely choose one's own signature creation mechanism, and consequently the level of assurance its messages carry. This causes an inherent asymmetry between the levels of trust one can have in the messages coming from Estonian and Finnish X-Road instances. eIDAS does not provide a mechanism for communicating the level of assurance required by the receiver of the signature. The only choice the receiver has is not to accept messages that carry a lower level of assurance, but this comes with a price of rejecting all the communication with the party that uses low-level certificates. This clearly contradicts the whole idea of X-Road federation.

It is also the case that existing international legislation (like eIDAS) covers only a fraction of all the required legal context. For example, dispute resolution must take place under some jurisdiction. In case of Estonian-Finnish federation, Estonian legal system is selected to be the primary dispute resolution context [5]. However, it is an open question what kind of a legal framework is appropriate for larger federations.

Another complicated legal issue is caused by possible incompatibility of certification policies in different jurisdictions. Even though on the technical level certificate interoperability can be rather well achieved by following the same standard (typically X.509v3 [2]), not all the aspects of trust are established using technical measures. Every Certification Authority also follows a certification policy that specifies a set of requirements and best practices that the CA follows when issuing the certificates.

A certification policy determines how certificate holders' identities are verified, how certificate life cycle is managed, how validity information is distributed, how all the processes are be audited, etc. A typical certification policy comprises of dozens of pages of loosely structured text. Making sure that two certification policies coming from different sources are in some sense compatible is a highly non-trivial task.

5   Technical Challenges

There is a number of technical parameters determining the service and trust level of X-Road infrastructure and messages exchanged over it. To ensure meaningful interoperability, these parameters must be comparable between different federated instances. Table 1 summarizes the parameter values in case of Estonian and Finnish deployments [3,4,10].

Table 1. Parameter comparison for Estonian and Finnish X-Road instances. The values marked with ∗ are not formally regulated

Parameter	Estonian instance	Finnish instance
Validity of OCSP responses	8 h	23.5 h
Validity of global configuration	6 h	72 h
Minimal required hash function	SHA-256	SHA-256*
Minimal required asymmetric algorithm	RSA 2048	RSA 2048*

Note that in the case of Finnish X-Road instance, the minimal required security level of cryptographic algorithms is not defined by a formal regulation, but by simply listing the corresponding TLS cipher suites as allowed as part of the security servers' configuration.

From Table 1 we see that the biggest discrepancy between the Estonian and Finnish instances is in the validity periods of OCSP responses and global configuration. In a regular day-to-day operation these differences should not matter much. However, the core idea of X-Road is to provide reliable data for decision making, with the option of holding the data source responsible if incorrect data can be proven to be the cause of an incorrect decision.

For example, in the case of OCSP responses it is possible that a certificate has been compromised and revoked, but one of the OCSP responses with a longer validity period can still be accepted by the members of one X-Road instance. Who should be held responsible in case a questionable decision has been made as a result is currently an open legal issue.

There are also other operational differences between the Estonian and Finnish X-Road instances. The original Estonian X-Road was built to support decision-making process based on the data obtained from other parties. In order to be able to later prove rightfulness of the decisions, the data needs to carry long-term evidentiary value. For this reason, all the X-Road messages are signed and timestamped.

Note that signing and time-stamping alone do not guarantee long-term proof value. The messages also need to be stored for later verification. This is why security servers in the Estonian X-Road instance support extensive message logging and archival.

In case of the Finnish instance, however, logging facilities are not utilized. This creates a potential situation where a Finnish X-Road member takes a decision based on the data obtained from, say, some Estonian collaborator, but will later be unable to prove the correctness of its actions. It is impossible to predict the outcome of the following disputes.

Another challenging aspect is authorization within member organizations. X-Road queries must be initiated by, and the results should eventually be used or interpreted by someone. X-Road infrastructure only deals with access control on the organization level. End-user authorization and access control within the service client's information systems are left as responsibilities of the service clients themselves. If these management practices are lax, an X-Road member sharing its data sets is risking a potential privacy leak due to a careless employee of the partner organization.

For an X-Road member, it is very hard to impose formal access control requirements on another organization, or verify that these requirements are fulfilled. This problem is even more serious concerning an organization in another country.

In principle, this problem should be solved by service use agreements – when gaining access to a service, the service client agrees to implement the required controls for authentication, access control and managing the received private data. The service provider opens the service only if it is satisfied with the level of security implemented by the client. However, for service providers with many clients, the case-by-case approach does not scale. It is infeasible to audit all the client information systems for compliance with the requirements. Thus, instead of treating each client separately, the service provider can require that all the clients implement a common security standard (assuming, of course, that the standard complies with the requirements of the service provider).

Both Estonia and Finland have established frameworks for assessing and ensuring security levels of governmental information systems. In case of Estonia, a baseline security system ISKE, a derivative of German BSI, is established[3]. Its Finnish counterpart is called VAHTI[4]. However, both frameworks are extensive (for example, if printed out, ISKE threat and countermeasure catalogues span over 3000 pages). To the best of our knowledge, there has been no thorough comparative analysis of the two, so from the viewpoint of a member of one X-Road instance, it is very hard to tell what security level can be assumed from a member of another instance.

It should be stressed again that these frameworks are only compulsory for governmental information systems. Private companies can implement these requirements, too, if they choose to, but there is no such obligation. Consequently, it is harder to state anything about the security level of data handling practices for private X-Road members.

6   Conclusions

Federating infrastructures like X-Road is inevitable in order to utilize data across national borders. However, there are still more question than there are answers.

---

[3] https://www.ria.ee/en/iske-en.html.
[4] https://www.vahtiohje.fi/web/guest.

Even though Estonia and Finland are close both geographically and culturally, there exist many differences in legislation, technical solutions and best practices. These differences have the potential to cause non-matching interpretations of various events, which in turn may lead to an unclear state of possible disputes. This contradicts the overall ideology of introducing X-Road in the first place.

This paper pointed out some of the most urgent problems that need to be addressed in both legal and technical aspects. However, even though there exist similar technical and operational standards in the two considered countries, not all the implementation aspects have been nor can be fully aligned.

While X-Road federation is still in the planning state, these issues are easier to fix than on the running system. On the other hand, actual severity of the identified problems is rather hard to assess without observing them in practice. Thus this line of research needs to be continued throughout the life cycle of federated X-Road implementation.

Acknowledgements. The research leading to these results has received funding from the European Regional Development Fund through Estonian Centre of Excellence in ICT Research (EXCITE) and the Estonian Research Council under Institutional Research Grant IUT27-1.

References

1. Regulation (EU) No 910/2014 of the European Parliament and of the Council of 23 July 2014 on electronic identification and trust services for electronic transactions in the internal market and repealing Directive 1999/93/EC. http://eur-lex.europa.eu/eli/reg/2014/910/oj
2. Internet X.509 Public Key Infrastructure Certificate and Certificate Revocation List (CRL) Profile (2008), iETF RFC 5280. https://tools.ietf.org/html/rfc5280
3. Infosüsteemide andmevahetuskihi Eesti keskkonna tingimused (2016). https://www.ria.ee/public/x_tee/Eesti_keskkonna_tehnilised_tingimused.pdf
4. Infosüsteemide andmevahetuskihi usaldusteenuste tingimused (2016). https://www.ria.ee/public/x_tee/usaldusteenuste_tingimused.pdf
5. Trust federation of Estonian X-tee and Finnish Palveluväylä. General Agreement between Estonian Information System Authority and Finnish Population Register Centre (2016). https://esuomi.fi/wp-content/uploads/2016/09/RIA_VRK_Trust_federation_Agreement_.pdf
6. Annuk, S., Nõgisto, I., Freudenthal, M., Mattila, J., Kallio, S.: X-Road: Protocol for Downloading Configuration (2017). https://github.com/ria-ee/X-Road/blob/develop/doc/Protocols/pr-gconf_x-road_protocol_for_downloading_configuration.md
7. Ansper, A., Buldas, A., Freudenthal, M., Willemson, J.: Scalable and efficient PKI for inter-organizational communication. In: Proceedings of 19th Annual Computer Security Applications Conference, pp. 308–318, December 2003
8. Freudenthal, M., Hanson, V., Nõgisto, I., Kromonov, I., Annuk, S., Seppälä, I.: X-Road Architecture. Technical Specification, version 1.3 (2015). https://github.com/vrk-kpa/xroad-public/tree/master/src/doc/Architecture

9. Kalja, A., Vallner, U.: Public e-Service Projects in Estonia. In: Proceedings of the Baltic Conference, BalticDB&IS 2002, vol. 2, pp. 143–154. Institute of Cybernetics at Tallinn Technical University (2002)
10. Kivimäki, P.: Trust federation of Estonian X-tee and Finnish Palveluväylä. General Agreement between Estonian Information System Authority and Finnish Population Register Centre. Annex X - Environments (2016). Version 0.1
11. Willemson, J., Ansper, A.: A secure and scalable infrastructure for inter-organizational data exchange and eGovernment applications. In: 2008 Third International Conference on Availability, Reliability and Security, pp. 572–577, March 2008

Strongly Deniable Identification Schemes Immune to Prover's and Verifier's Ephemeral Leakage

Łukasz Krzywiecki$^{(\boxtimes)}$ and Marcin Słowik

Department of Computer Science, Faculty of Fundamental Problems of Technology,
Wrocław University of Science and Technology, Wrocław, Poland
{lukasz.krzywiecki,marcin.slowik}@pwr.edu.pl

Abstract. In this paper, we consider *Identification Schemes* (IS) in the context of attacks against their deniability via Fiat-Shamir transformations. We address the following issue: How to design and implement a deniable IS, that is secure against ephemeral leakage on both a Prover's and a Verifier's side, and withstands attacks based on Fiat-Shamir transformation. We propose a new security model to address the leakage on the Verifier's side, extending the previous propositions [1]. During the *Query Stage*, we allow the malicious Verifier to set random values used on the Prover's side. Additionally, we allow malicious Prover to access ephemeral values of the Verifier during the *Impersonation Stage*. We introduce two generic constructions based on three-step IS. Finally, we provide an example scheme based on the extended construction from [1], which is provably deniable and secure in our new strong model.

Keywords: Identification scheme · Ephemeral secret setting · Ephemeral secret leakage · Deniability · Simulatability · Zero-knowledge proofs

1 Introduction

Identification schemes (IS) based on *Public Key Infrastructure* (PKI) allow a Prover, holding a secret key, to prove its possession via a zero-knowledge protocol executed with a Verifier holding a corresponding public key. There are two common requirements that IS should satisfy: (1) *security* - a malicious Prover should not be able to successfully complete the protocol without the corresponding secret key; (2) *privacy* - in some scenarios, the protocol should be deniable, meaning that its transcript must not be a strong proof of Prover's participation. Alternatively, there are cases in which the protocol should not be deniable and must provide a strong proof of Prover's participation. Typically, ISes require complex computations over large numbers, and are deployed on the users' electronic

---

Partially supported by funding from Polish National Science Centre (NCN) contract number DEC-2013/08/M/ST6/00928.

© Springer International Publishing AG 2017
P. Farshim and E. Simion (Eds.): SecITC 2017, LNCS 10543, pp. 115–128, 2017.
https://doi.org/10.1007/978-3-319-69284-5_9

devices, which store sensitive secret keys. There are several common threats concerning this aspect, emerging from the fact that the end users see the devices as *black boxes*, and they have to trust that the scheme implementation processes are not tampered with. Very often, such devices are produced by vendors beyond of the end users' control, and as such are subject to malicious modification, which can bring about the following vulnerabilities:

- **Prover's Ephemeral Leakage**: Especially important for three round identification schemes, with three messages exchanged between a *Prover* and a *Verifier*:
  (1) the *Prover* sends a *commitment* to a random value to the *Verifier*;
  (2) the *Verifier* sends to the Prover another random value called a *challenge*;
  (3) a *response* message sent by the Prover is a result of a function of the challenge and the secret key masked by the committed ephemeral.
  At the Verifier's side, this *response* is checked by the means of the public key with the commitment and the challenge. If a malicious manufacturer implements a covert channel within a Prover's device, it can learn (or set) ephemeral values coined in the commitment phase, and unmask the secret key from the response. This way, the ephemeral leakage subsequently enables impersonation attacks using the Prover's identity. Note that Schnorr [2] and Okamoto [3] ISes are vulnerable to this attack. Recently, a remedy for that problem has been proposed in [1]. The solution is quite flexible and works for many similar three round constructions.
- **Verifier's Ephemeral Leakage**: Alternatively, if there is a back-door channel in a Verifier's device, it can be exploited by a malicious Prover to read ephemeral values coined by the Verifier before the challenge phase. There are ISes which rely on the secrecy of such values e.g. [4–6]. In all these schemes the Adversary knowing the Verifier's ephemeral value can impersonate the Prover without the secret key. It is worth to notice that typical three round identification schemes are immune, from their design, to attacks based on the Verifier's ephemeral leakage, since the only random value of the Verifier is the challenge revealed to the Prover in the second message. This statement, however, requires an assumption that the challenge value is coined strictly *after* the *commitment* phase, as otherwise impersonation would be trivial, due to simulatability property of the IS.
- **Losing Deniability**: Although typical three round ISes resist Verifier's ephemeral leakage attacks, they suffer from the deniability attacks mounted by the active malicious Verifier. Indeed, instead of coining the challenge at random, the Adversary can use a Fiat-Shamir transformation [7] and compute challenge as a hash value over the commitment, this way changing the scheme into an undeniable signature.

Problem Statement: In this paper we address the following issue: *How to design and implement a deniable* IS:

(1) *secure against ephemeral leakage on both Prover's and Verifier's side;*
(2) *withstanding attacks based on Fiat-Shamir transformation.*

1.1   Contribution of the Paper

The contribution of the paper is the following:

- We introduce a new strong security model for deniable identification schemes in which we allow *Adversaries*:
  - to set ephemerals on Provers' side in the Query Stage of the security experiment,
  - to read ephemerals used on Verifiers' side in the final (Impersonation) Stage of the security experiment.

    We define the IS to be secure if no Adversary, even given such a power and knowledge, is able to impersonate a Prover, without their secret key.
- We propose a general extension to three-rounds identification protocols, e.g. [1–3], hardening them against *Attacks on Deniability* by Fiat-Shamir transformation, secure in our stronger model.
- We show an example of our extension based on a modified Schnorr scheme, and prove its security in our model.

Our proposition is useful for systems based on three-round IS, where randomness leakage is possible. There is a growing demand for schemes secure in such scenarios, due to recent revelations regarding undermining cryptographic standards and implementations.

Remark: note that typical, 4-round *Malicious Verifier Zero Knowledge* schemes, that are based on commitments to challenge are not secure in the *Verifier Leakage* model. Coining challenge before Prover's commitment is sent may lead to straightforward impersonation: the challenge leakage allows for textbook simulation.

Previous Work. Identification schemes have been in use since the dawn of the modern, public-key cryptography [2,7–9]. Schnorr has introduced a DLP based construction [2], followed by [3] of Okamoto. Several ISes are specialized in terms of models or attack schemes, e.g. [10,11]. [12] introduced a notion of vulnerability to ephemeral leakage and proposed IS protocols invulnerable to such attacks. [13] shown IS secure against *Reset Attacks* based on stateless, deterministic signature schemes, CCA-secure asymmetric encryption schemes and pseudorandom functions with trapdoor commitments. *Subversion resilience* is a concept regarding security of various schemes in settings, where malicious manufacturer may replace original scheme with a modified one that behaves identically, but may leak additional information by hidden trapdoors in regular outputs [14–16].

The paper is organized in the following way. In Sect. 2 we review our strong security model, strongly based on models from [1]. In Sect. 3 we propose the extensions of generic three-rounds ISes following the *commit, challenge, response* schema, which protects against Fiat-Shamir transformation-based attacks on deniability. In Sect. 4 we modify the protocol from [1], and prove its security in our model.

2   System Model

Let us first recall the definition of IS from [1] loosely based on Okamoto's definition [3].

Definition 1 (Identification Scheme). *An identification scheme* IS *is a tuple of procedures* $(\mathsf{PG}, \mathsf{KG}_\mathcal{P}, \mathsf{KG}_\mathcal{V}, \mathcal{P}, \mathcal{V}, \pi)$:

par $\leftarrow$ PG(1^λ): *takes the parameter* λ, *and outputs public parameters.*
$(\mathsf{sk}, \mathsf{pk}) \leftarrow \mathsf{KG}_\mathcal{P}(\mathsf{par})$: *outputs secret and public keys of the prover.*
$(\mathsf{se}, \mathsf{pe}) \leftarrow \mathsf{KG}_\mathcal{V}(\mathsf{par})$: *(optional) outputs secret and public keys of the verifier.*
$\mathcal{P}(\mathsf{sk}, \mathsf{pe})$: *denotes the Prover algorithm which interacts with the Verifier* $\mathcal{V}$.
$\mathcal{V}(\mathsf{pk}, \mathsf{se})$: *denotes the Verifier algorithm which interacts with the Prover* $\mathcal{P}$.
$\pi(\mathcal{P}, \mathcal{V})$: *denotes the protocol of interactions between* $\mathcal{P}$ *and* $\mathcal{V}$.

IS *has Initialization and Operation Stages. In Initialization Stage, parameters and keys for users are generated. In the latter, a user proves interactively its identity in front of the Verifier:* $\pi(\mathcal{P}(\mathsf{sk}, \mathsf{pe}), \mathcal{V}(\mathsf{pk}, \mathsf{se}))$. *We write* $\pi(\mathcal{P}, \mathcal{V}) \to 1$ *if* $\mathcal{P}$ *and* $\mathcal{V}$ *have mutually accepted each other in* π. *The scheme is* complete *iff*

$$\Pr[(\mathsf{sk}, \mathsf{pk}) \leftarrow \mathsf{KG}_\mathcal{P}(), (\mathsf{se}, \mathsf{pe}) \leftarrow \mathsf{KG}_\mathcal{V}(), \pi(\mathcal{P}(\mathsf{sk}, \mathsf{pe}), \mathcal{V}(\mathsf{pk}, \mathsf{se})) \to 1] = 1.$$

The optional, verifier key pair $(\mathsf{se}, \mathsf{pe})$ exists in several IS schemes. If the IS does not rely on it, or even explicitly denies its existence, we may assume $\mathsf{KG}_\mathcal{V}$ always returns $(\bot, \bot)$ on any input.

2.1   Impersonation Resilience

The fundamental security requirement for IS is that no malicious Prover algorithm $\mathcal{A}$, without the secret key sk corresponding to the public key pk used by the Verifier, should be accepted in protocol π. In other words, we require that probability $\Pr[\pi(\mathcal{A}(\mathsf{pk}, \mathsf{pe}), \mathcal{V}(\mathsf{pk}, \mathsf{se})) \to 1] \leq \epsilon_\lambda$ where ϵ_λ is a negligible function. We formally define our security model in Sect. 2.3.

2.2   Adversary Model

The process in which an Adversary gains knowledge about the attacked protocol is modeled by a *Query Stage* of the security experiment. This means that the Adversary runs a polynomial number ℓ of the protocol executions between the Prover and the Verifier: $\pi(\mathcal{P}(\mathsf{sk}, \mathsf{pe}), \mathcal{V}(\mathsf{pk}, \mathsf{se}))$. We consider the *Active Adversary* which actively participates in the stage, usually as a Verifier $\widetilde{\mathcal{V}}$, i.e. it actively chooses messages sent to the Prover. Based on [1], we assume the Adversary additionally adaptively sets the ephemeral values for the Prover in each protocol run in the *Query Stage*. Finally, extending the model from [1], we consider the Adversary that can read ephemeral values of the Verifier in the *Impersonation Stage*, immediately after those values are produced.

2.3   Security Experiments

Let $\bar{x}_i$ be adaptive ephemerals from a malicious Verifier $\widetilde{\mathcal{V}}$ injected to the Prover $\mathcal{P}^{\bar{x}_i}$ in the ith execution of the *Query Stage*. Let the view $v_i = \{T_1, \ldots, T_i\} \cup \{\bar{x}_1, \ldots, \bar{x}_i\}$ be the total knowledge $\mathcal{A}$ can gain after i runs of π, where T_i is the transcript of the protocol messages in the ith execution. The IS is CPLVE-secure if such a cumulated knowledge after ℓ executions does not help the Adversary to be accepted by the Verifier except with a negligible probability.

Definition 2 (Chosen Prover-Leaked Verifier Ephemeral – (CPLVE)).
Let IS = (PG, $KG_{\mathcal{P}}$, $KG_{\mathcal{V}}$, $\mathcal{P}$, $\mathcal{V}$, π). *We define security experiment* $\mathrm{Exp}_{IS}^{CPLVE,\lambda,\ell}$:

Init Stage: par $\leftarrow$ PG(1^λ), (sk, pk) $\leftarrow$ $KG_{\mathcal{P}}$(par), (se, pe) $\leftarrow$ $KG_{\mathcal{V}}$(par).
$\quad$ $\mathcal{A}$:($\widetilde{\mathcal{P}}$(pk, pe), $\widetilde{\mathcal{V}}$(pk, pe)).
Query Stage: *For $i = 1$ to ℓ run* $\pi(\mathcal{P}^{\bar{x}_i}$(sk, pe), $\widetilde{\mathcal{V}}$(pk, pe, $\bar{x}_i$, v_{i-1}))*, where* $\bar{x}_i \in \{\bar{x}_1, \ldots, \bar{x}_\ell\}$ *are the adaptive ephemerals from* $\widetilde{\mathcal{V}}$ *injected to the Prover* $\mathcal{P}^{\bar{x}_i}$ *in the ith execution, and v_{i-1} is the total view of $\mathcal{A}$ until the ith execution.*
Impersonation Stage: $\mathcal{A}$ *executes the protocol* $\pi(\widetilde{\mathcal{P}}$(pk, pe, v_ℓ, $\bar{e}$), $\mathcal{V}$(pk, se))*, where $\bar{e}$ are the ephemerals of the Verifier leaked to the malicious Prover $\widetilde{\mathcal{P}}$.*

The advantage of $\mathcal{A}$ in the experiment $\mathrm{Exp}_{IS}^{CPLVE,\lambda,\ell}$ *is the probability of acceptance in the last stage:*

$$\mathbf{Adv}(\mathcal{A}, \mathrm{Exp}_{IS}^{CPLVE,\lambda,\ell}) = \Pr[\pi(\widetilde{\mathcal{P}}(\mathsf{pk}, \mathsf{pe}, v_\ell, \bar{e}), \mathcal{V}(\mathsf{pk}, \mathsf{se})) \rightarrow 1].$$

We say that the IS *is* (λ, ℓ)-CPLVE–*secure if* $\mathbf{Adv}(\mathcal{A}, \mathrm{Exp}_{IS}^{CPLVE,\lambda,\ell}) \leq \epsilon_\lambda$ *and ϵ_λ is negligible in λ.*

We utilize the definition of deniability from [17], which itself generalizes the idea from [18]. Let π be a protocol in IS. We assume an adversary $\mathcal{M}$ which inputs an arbitrary number of public keys $\mathbf{pk} = (\mathsf{pk}_1, \ldots, \mathsf{pk}_\ell)$, randomly coined with an appropriate key generating algorithm. The adversary initiates an arbitrary number of protocols with the honest parties, some in a role of the prover, others in a role of the verifier. The view of $\mathcal{M}$ consists of its internal randomness, and the transcript of the entire interaction, in all the protocols in which $\mathcal{M}$ participated. We denote this view as $\mathrm{View}_{\mathcal{M}}(\mathbf{pk}, a)$.

Definition 3. *We say that π is a strongly deniable protocol of* IS *with respect to the class A of auxiliary inputs if for any adversary $\mathcal{M}$, for any input of public keys $\mathbf{pk} = (\mathsf{pk}_1, \ldots, \mathsf{pk}_\ell)$ and any auxiliary input $a \in A$, there exists a simulator $SIM_{\mathcal{M}}$ that, running on the same inputs as $\mathcal{M}$, produces a simulated view which is indistinguishable from the real view of $\mathcal{M}$. That is, consider the following two probability distributions, where $\mathbf{pk} = (\mathsf{pk}_1, \ldots, \mathsf{pk}_\ell)$ is the set of public keys of the honest parties:*

$$\mathcal{R}eal(\lambda, a) = [(\mathsf{sk}_i, \mathsf{pk}_i) \leftarrow KG(1^\lambda); (a, \mathbf{pk}, \mathrm{View}_{\mathcal{M}}(\mathbf{pk}, a)]$$
$$\mathcal{S}im(\lambda, a) = [(\mathsf{sk}_i, \mathsf{pk}_i) \leftarrow KG(1^\lambda); (a, \mathbf{pk}, SIM_{\mathcal{M}}(\mathbf{pk}, a)]$$

then for all probabilistic poly-time machines Dist *and all* $a \in A$, *there exists a function* ϵ_λ *negligible in* λ *s.t.:*

$$|\Pr{}_{x \in \mathcal{R}eal(\lambda,a)}[\text{Dist}(x) = 1]| - |\Pr{}_{x \in \mathcal{S}im(\lambda,a)}[\text{Dist}(x) = 1]| \le \epsilon_\lambda.$$

The idea behind this definition is that no adversary can follow a strategy that is not simulatable, i.e. there exist a distinguisher differentiating between the real adversary and a simulator. In other words, all adversarial strategies are simulatable.

2.4   Deniability Attack in Active Mode

Let $T = (X, c, S)$ denote the transcript of a 3-round IS. In Fig. 1 we recall how active Verifier can use the Fiat-Shamir transformation to generate undeniable transcript of the protocol, effectively transforming the 3-round interactive IS into non-interactive signature scheme. The value r is a randomizing factor. In real signature schemes, the value r is replaced by message m. The hash input $i = (X, r)$ is an undeniable proof that the party $\mathcal{P}$ has participated in the protocol.

Let $\text{IS} = (\text{PG}, \text{KG}_\mathcal{P}, \mathcal{P}, \mathcal{V}, \pi)$ be a secure 3-round identification scheme, where (X, c, S) denotes commitment, challenge, and response messages. Let (sk, pk) denote a pair of secret/public keys. Let $\mathcal{H}$ denote a secure collision resistant one way hash function into the challenge space. The *deniability breaking attack* on protocol $\pi(\mathcal{P}(\text{sk}), \mathcal{V}(\text{pk}))$:

1. $\mathcal{P}$: prepare X, $\mathcal{P} \overset{X}{\rightsquigarrow} \mathcal{V}$.
2. $\mathcal{V} : r \leftarrow_R \{0, 1\}^n, c = \mathcal{H}(X, r)$, $\mathcal{P} \overset{c}{\leftsquigarrow} \mathcal{V}$.
3. $\mathcal{P}$: prepare S, $\mathcal{P} \overset{S}{\rightsquigarrow} \mathcal{V}$.
4. $\mathcal{V}$: verify in a regular way.

Fig. 1. The attack on deniability of typical 3-round IS.

3   Extended Identification Schemes

3.1   General Idea – Commitment to an Unknown Value

The general idea behind the proposed extensions is that in order to achieve the strong deniability property in the *Verifier Ephemeral Leakage* scenario, the Verifier has to prove that the challenge has not been produced via the transformation of the Prover's commitment X. Therefore, at the beginning of the protocol, the Verifier itself randomly chooses a commitment to an unknown challenge, which can be opened by them only after they obtain the first message from the Prover. We propose two different methods for this purpose: (a) *Deterministic Encryption Method*; (b) *Proof of Computation Method*; which can be used separately or together.

3.2   Deterministic Encryption Method

This extension is based on the assumption that the scheme in subject can be used in conjunction with a deterministic asymmetric encryption, for which, w.l.o.g., we use the following definition.

Definition 4 (Asymmetric Encryption Scheme). *Let* $E = (KG_E, \mathcal{E}, \mathcal{D})$ *denote a secure deterministic encryption scheme, s.t.* $(se, pe) \leftarrow KG_E()$:

(1) $\forall_{(m \in M)} : \mathcal{E}(pe, m) \to c \in C, s.t. \mathcal{D}(se, c) \to m,$
(2) $\forall_{(c \in C)} : \mathcal{D}(se, c) \to m \in M, s.t. \mathcal{E}(pe, m) \to c$

where (se, pe) *is a secret/public key pair;* M, C *are plaintext, and ciphertext spaces;* $(KG_E, \mathcal{E}, \mathcal{D})$ *are key generation, encryption and decryption algorithms.*

The only security property of E that is required in the proposed scheme is its *secrecy* or *one-wayness*, that is:

Definition 5 (Encryption One-Wayness). *An Asymmetric Encryption Scheme* E *has encryption one-wayness property, if for any PPT algorithm* $\mathcal{A}$, *for* $(se, pe) \leftarrow KG_E(1^\lambda)$ *and for a* $c \in C$ *selected uniformly at random:*

$$\Pr[\mathcal{A}(pe, c) = \mathcal{D}(se, c)] \leq \epsilon_\lambda$$

for a negligible function ϵ_λ.

Note that the equation is actually equivalent to $\Pr[\mathcal{E}(pe, \mathcal{A}(pe, c)) = c] \leq \epsilon_\lambda$ and to $\Pr[\mathcal{A}(pe, \mathcal{E}(pe, m)) = m] \leq \epsilon_\lambda$ for uniformly selected message $m \in M$.

An example of such a scheme is a textbook RSA Encryption [19]. With the E scheme, as of Definition 4, the extension is the following: at the beginning the Verifier chooses the ciphertext $\hat{c}$ randomly, which is immediately sent to the Prover. This is a commitment to a yet unknown challenge c, and corresponds to the Verifier's ephemeral value, known to the malicious Prover in the *Verifier Ephemeral Leakage* model. Then, the Verifier waits until it gets a commitment from the Prover and only then opens the commitment $m = \mathcal{D}(se, \hat{c})$, chooses a random bit $b \leftarrow_R \{0, 1\}$ and sends m, b to the Prover. The bit b allows for randomization of c, but the information size of b is insufficient to indicate the Prover's identity, as both options are equally simulatable. Both parties compute the commitment with a secure one way hash function $c = \mathcal{H}(m, b)$. This reflects the situation in which both the Prover and the Verifier learn the commitment c only after X has been received by the Verifier. On the other hand, the Prover checks if the value m agrees with the commitment $\mathcal{E}(pe, m) \stackrel{?}{=} \hat{c}$, and then it is convinced that the challenge m has not been produced by a Fiat-Shamir-like transformation over its own commitment X. If $\mathcal{E}(pe, m) \neq \hat{c}$, the Prover stops the protocol.

The proposed extension is depicted in Fig. 2. Note that the IS has a slightly different interface as $\mathcal{P}$ and $\mathcal{V}$ take each others' public keys and their own secret keys on input (contradictory to the Definition 1 where only Prover's keys were considered). The single random bit b has a very small influence on the protocol

Let $\mathsf{IS} = (\mathsf{PG}, \mathsf{KG}_{\mathcal{P}}, \mathcal{P}, \mathcal{V}, \pi)$ be a secure 3-round identification scheme, where (X, c, S) denotes commitment, challenge, and response messages. Let $\mathcal{H}$ denote a secure collision resistant one way hash function into the challenge space. Let $(\mathsf{KG}_{\mathsf{E}}, \mathcal{E}, \mathcal{D})$ denote a secure encryption scheme. We add $\mathsf{KG}_{\mathcal{V}} = \mathsf{KG}_{\mathsf{E}}$ and modify the protocol $\pi(\mathcal{P}(\mathsf{sk}, \mathsf{pe}), \mathcal{V}(\mathsf{pk}, \mathsf{se}))$ in the following way:

0. $\mathcal{V}$: commits to unknown challenge $\hat{c} \in_R C$, $\mathcal{P} \xleftarrow{\hat{c}} \mathcal{V}.$
1. $\mathcal{P}$: prepares X, $\mathcal{P} \xrightarrow{X} \mathcal{V}.$
2. $\mathcal{V}$: waits for X, $m = \mathcal{D}(\mathsf{se}, \hat{c})$, $b \leftarrow_R \{0, 1\}$, $\mathcal{P} \xleftarrow{m,b} \mathcal{V}.$
3. $\mathcal{P}$: if $(\mathcal{E}(\mathsf{pe}, m) \stackrel{?}{=} \hat{c})$: $c = \mathcal{H}(m, b)$, prepares S, $\mathcal{P} \xrightarrow{S} \mathcal{V}.$
4. $\mathcal{V}$: $c = \mathcal{H}(m, b)$, verifies (X, c, S) in a regular way.

Fig. 2. Extension based on encryption scheme.

itself, but is crucial in proving the security of the underlying IS, when the proof uses *rewinding techniques*, in order to produce two distinct challenges for the same initial commitment.

Lemma 1. *The extension proposed in Fig. 2 protects against deniability attacks on 3-round IS via Fiat-Shamir transformation - as of Fig. 1.*

Proof. The proof is by contradiction. Assume that a malicious Verifier successfully, with non-negligible probability, mounts the attack resulting with transcript $T = (\hat{c}, X, m, b, S)$ and the proof $i = (X, r)$, s.t.: $m = \mathcal{D}(\mathsf{se}, \hat{c})$, $c = \mathcal{H}(m, b)$ and $c = \mathcal{H}'(X, r)$ for any hash function $\mathcal{H}'$, then we successfully find a collision for the hash function $\mathcal{H}$ with inputs $i = (m, b)$ and $i = (X, r)$ (if $\mathcal{H} = \mathcal{H}'$), or break preimage resistance of either $\mathcal{H}$ (with the image being $c = \mathcal{H}'(X, r)$) or $\mathcal{H}'$ (with the image being $c = \mathcal{H}(m, b)$). □

Lemma 2. *The extension proposed in Fig. 2 retains zero-knowledge properties of the underlying IS.*

Proof (Sketch).

Completeness. Straightforward verification shows that if the original IS was complete, the modified scheme is complete as well. The addition of $\hat{c}$ and the way c is computed does not influence the protocol if only $\mathcal{H}$ is a secure hash function indistinguishable from a Random Oracle into the challenge space.

Soundness. The method of proving soundness of the modified scheme is closely related to the method used to prove the soundness of IS. In principle, $\mathcal{P}$ cannot derive any knowledge from the commitment scheme except with a negligible probability. If $\mathcal{P}$ could derive any information about the challenge message before the *commitment* phase, they would be able to break the *encryption one-wayness* of E' (cf. Definition 5).

Zero-knowledge. The protocol is simulatable if only IS is simulatable. Let us choose $m \in M$ and $b \in \{0,1\}$ at random. Compute $c = \mathcal{H}(m,b)$ and simulate transcript (X,c,S) of IS for the given challenge c. Compute commitment $\hat{c} = \mathcal{E}(\mathsf{pe},m)$. Return $(\hat{c}, X, (m,b), S)$ as the simulated transcript.   □

3.3   Proof of Computation Method

This extension is based on the assumption that the Verifier's computing device $\mathsf{D}_\mathcal{V}$ is faster than the Prover's computing device $\mathsf{D}_\mathcal{P}$. Let $\mathsf{RT}_\mathsf{D}(A)$ denote a running time of the device D executing an algorithm A. Let (P, X) denote a computational problem in domain X, and ς denote its solution. Let $\mathsf{Ver}(\mathsf{P}, X, \varsigma)$ denote a fast verification algorithm which returns 1 if ς is a solution for (P, X) or returns 0 otherwise. Let $\mathcal{S}(\mathsf{P}, X)$ denote the algorithm solving (P, X). We assume that $\mathcal{S}(\mathsf{P}, X)$ is "quite" complex, that is, on any device D it holds that: $\mathsf{RT}_\mathsf{D}(\varsigma = \mathcal{S}(\mathsf{P}, X)) \gg \mathsf{RT}_\mathsf{D}(\mathsf{Ver}(\mathsf{P}, X, \varsigma))$. To capture that the Verifier's computing device $\mathsf{D}_\mathcal{V}$ is faster than the Prover's computing device $\mathsf{D}_\mathcal{P}$ we assume that: $\mathsf{RT}_{\mathsf{D}_\mathcal{V}}(\mathcal{S}(\mathsf{P}, X)) < \mathsf{RT}_{\mathsf{D}_\mathcal{P}}(\mathcal{S}(\mathsf{P}, X))$, for any $(\mathsf{P}, X, \mathcal{S})$.

Let $\mathcal{G}(\mathsf{P}, w)$ be a domain generation algorithm for problem P that takes a seed $w \in Seed$ as an input, and outputs a domain X for P. Let $\mathcal{H} : \{0,1\}^* \to Seed$ be a one way function used to compute a seed w for $\mathcal{G}(P, w)$. Assume the following process of generating a sequence of problems P, X_i and its solutions ς_i from the random seed $w \in_R Seed$.

$\mathsf{Gen}(\mathsf{P}, w)$:
Init Stage: $n = 0$, $X_0 = \mathcal{G}(\mathsf{P}, w)$, $\varsigma_0 = \mathcal{S}(\mathsf{P}, X_0)$
Iterate since Start signal until Stop signal:
$n = n + 1$, $w_n = \mathcal{H}(\varsigma_{n-1})$, $X_n = \mathcal{G}(\mathsf{P}, w_n)$, $\varsigma_n = \mathcal{S}(\mathsf{P}, X_n)$,
Return: $\langle \varsigma_i \rangle_i^n$

Assume the verification process:

$\mathsf{Check}(\mathsf{P}, w, \langle \varsigma_i \rangle_i^n)$:
Init Stage: $n = 0$, $X_0 = \mathcal{G}(\mathsf{P}, w)$, $v_0 = \mathsf{Ver}(\mathsf{P}, X_0, \varsigma_0)$
Iterate for all $i \in \{1 \ldots n\}$: $w_i = \mathcal{H}(\varsigma_{i-1})$, $X_i = \mathcal{G}(\mathsf{P}, w_i)$, $v_i = \mathsf{Ver}(\mathsf{P}, X_i, \varsigma_i)$
Return: $\prod_{i=0}^{n} v_i$

The Proof of Computation System PCS is a tuple of the above defined algorithms: $(\mathcal{G}, \mathsf{P}, \mathcal{S}, \mathsf{Ver}, \mathsf{Gen}, \mathsf{Check}, \mathcal{H})$. The proposed extension is depicted in Fig. 3

Lemma 3 *The extension proposed in Fig. 3 protects against deniability attacks on 3-round IS via Fiat-Shamir transformation - as of Fig. 1.*

Proof The proof is by contradiction. Similarly as in the proof of Lemma 1, if a malicious Verifier successfully, with non-negligible probability, attacks the

Let $IS = (PG, KG_{\mathcal{P}}, \mathcal{P}, \mathcal{V}, \pi)$ be a secure 3-round identification scheme, where (X, c, S) denote commitment, challenge, and response messages. Let $\mathcal{H}$ denote a secure one way hash function into the challenge space. Let $PCS = (\mathcal{G}, P, \mathcal{S}, Ver, Gen, Check)$ denote a proof of computation system. We modify the protocol $\pi(\mathcal{P}(\text{sk}), \mathcal{V}(\text{pk}))$ in the following way:

0. $\mathcal{V} : w \in_R Seed$, Start iterating $Gen(P, w)$ $\mathcal{P} \xleftarrow{\;w\;} \mathcal{V}.$

1. $\mathcal{P}$: prepares X, $\mathcal{P} \xrightarrow{\;X\;} \mathcal{V}.$

2. $\mathcal{V}$: waits for $X, \langle \varsigma_i \rangle_i^n =$ Stop iterating Gen, $\mathcal{P} \xleftarrow{\langle \varsigma_i \rangle_i^n} \mathcal{V}.$

3. $\mathcal{P}$: if $(\text{Check}(P, w, \langle \varsigma_i \rangle_i^n) \overset{?}{=} 1)$: $c = \mathcal{H}(\langle \varsigma_i \rangle_i^n)$, prepares S, $\mathcal{P} \xrightarrow{\;S\;} \mathcal{V}.$

4. $\mathcal{V}$: verifies (X, c, S) in a regular way.

Fig. 3. Extension based on Proof of Computation System.

scheme getting the transcript $T = (w, X, \langle \varsigma_i \rangle_i^n, S)$ and the Fiat-Shamir undeniability proof $i = (X, m)$, s.t: $\langle \varsigma_i \rangle_i^n = \text{Gen}(P, w)$, $c = \mathcal{H}(\langle \varsigma_i \rangle_i^n)$, and $c = \mathcal{H}(X, m)$, then we successfully find a collision for the hash function $\mathcal{H}$ with inputs $i = (\langle \varsigma_i \rangle_i^n)$ and $i = (X, r)$ (if $\mathcal{H} = \mathcal{H}'$), or break preimage resistance of either $\mathcal{H}$ (with the image being $c = \mathcal{H}'(X, r)$) or $\mathcal{H}'$ (with the image being $c = \mathcal{H}(\langle \varsigma_i \rangle_i^n)$). $\square$

4 Specific Scheme Proposition

To show the applicability of our propositions we introduce the modification of the scheme from [1] augmented with our first extension, using textbook RSA encryption. The proposed scheme is depicted in Fig. 4.

4.1 Simulation in the *Passive Adversary* Mode

The modified Schnorr IS preserves the simulatability property of its original version. The protocol transcript can be efficiently simulated by the following algorithm (for any public keys (pk, pe) and challenge message (m, b)):

Sim $S_{IS}^{PA}((\text{pk}, \text{pe} = (e, N)), (m, b))$:
     $\hat{c} = m^e$, $c = \mathcal{H}_q(m, b)$, $s \leftarrow_R \mathbb{Z}_q^*$,
     $X := (g^s / \text{pk}^c)$, $\hat{g} := \mathcal{H}_G(X, c)$, $S := \hat{g}^s$
     return:
         $T = (\hat{c}, X, (m, b), S)$

Observe that for this transcript the verification holds: $\hat{e}(S, g) = \hat{e}(\mathcal{H}_G(X, c), X\text{pk}^c)$. The simulator can play the simulated transcript $T = (\hat{c}, X, m, S)$ in the correct order, thus mimicking the real interaction between the parties. The real transcript and the simulated tuple are identically distributed.

---

par $\leftarrow$ PG(1^λ): Let $G, G_T \leftarrow \mathcal{G}(1^\lambda)$ be cyclic groups of a prime order q, and $g \in G$ be a generator. Let $\mathcal{H}_q : \{0,1\}^* \to \mathbb{Z}_q^*$ and $\mathcal{H}_G : \{0,1\}^* \to G$ be hash functions. Let $\hat{e} : G \times G \to G_T$ be a bilinear map. We assume that GDH holds in G, and $\hat{e}$ is $\mathcal{O}_{\mathsf{DDH}}$ oracle. Set par $= (q, G, g, G_T, \mathcal{H}_q, \mathcal{H}_G, \hat{e})$.

KG$_{\mathcal{P}}$(par): sk $\leftarrow_R \mathbb{Z}_q^*$, pk $= g^{\mathsf{sk}}$. Output (sk, pk).

KG$_{\mathcal{V}}$(par): Randomly select large primes a and b s.t. $N = ab$ satisfies λ bits of security; Select relatively small value $e \approx 2^{16}$ co-prime to the order of group $\mathbb{Z}_N^\times$ and compute its inverse d modulo the order. Output (se $= d$, pe $= (N, e)$).

$\pi(\mathcal{P}(\mathsf{sk}, \mathsf{pe}), \mathcal{V}(\mathsf{pk}, \mathsf{se}))$ is the following protocol:

0. $\mathcal{V}$: commits to an unknown challenge $\hat{c} \in_R \mathbb{Z}_N^\times$, $\mathcal{P} \xleftarrow{\hat{c}} \mathcal{V}$.

1. $\mathcal{P}$: selects $x \in_R \mathbb{Z}_q^*$, $X = g^x$, $\mathcal{P} \xrightarrow{X} \mathcal{V}$.

2. $\mathcal{V}$: awaits X, $m = \hat{c}^d$, $b \leftarrow_R \{0,1\}$, $\mathcal{P} \xleftarrow{m,b} \mathcal{V}$.

3. $\mathcal{P}$: aborts if $m^e \neq \hat{c}$, otherwise:
    $c = \mathcal{H}_q(m,b), \hat{g} = \mathcal{H}_G(X,c), S = \hat{g}^{x+\mathsf{sk}\cdot c}, \mathcal{P} \xrightarrow{S} \mathcal{V}$.

4. $\mathcal{V}$: $c = \mathcal{H}_q(m,b), \hat{g} = \mathcal{H}_G(X,c)$,
    accepts iff $\hat{e}(S,g)=\hat{e}(\hat{g}, X\mathsf{pk}^c)$.

Fig. 4. The proposed modified IS.

4.2 Security Analysis

In our analysis we assume that there is an effective Adversary that breaks our scheme from Fig. 4. In the Query Stage, we interact with the Adversary, simulating the proofs without the secret key, but using the injected ephemerals. In the Impersonation Stage, there are two mutually exclusive possibilities: either the Adversary knows the challenge $c = \mathcal{H}_q(m,b)$ before sending X, or he does not. Therefore, in our reduction proof, we guess in which alternative the Adversary exists. If it knows the value $c = \mathcal{H}_q(m,b)$, we use it to break underlying security of RSA. If the Adversary attacks without the knowledge of the challenge $c = \mathcal{H}_q(m,b)$ we proceed as in the original proof from [1]. In the latter case, we follow the methodology from [2,3], using *rewinding technique*. Namely, we fix randomness $\hat{c}, X$, but change the bit b by setting it to 0 for the first run, and to 1 for the second run. This results with two tuples $(\hat{c}, X, m, 0, S_1), (\hat{c}, X, m, 1, S_2)$ letting us solve the underlying hard problem – in this case CDH.

Theorem 4. *Let* IS *denote the modified identification scheme (as in Fig. 4).* IS *is secure (in the sense of Definition 2), i.e. the advantage* $\mathbf{Adv}(\mathcal{A}, \mathrm{Exp}^{\mathsf{CPLVE},\lambda,\ell})$ *is negligible in* λ*, for any PPT algorithm* $\mathcal{A}$*.*

We postpone the proof to the Appendix A.

5 Conclusion

In this paper, we have shown how to modify a wide class of three-move identification schemes secure against *Prover Ephemeral Injection* into identification schemes secure against *Verifier Ephemeral Leakage* and *Deniability Attack*. We have shown an example based on a modified Schnorr IS from [1]. We have formalized a security model and proved the security of our constructions.

A   Postponed Proof

Proof. We use ROM for hash queries. The proof is by contradiction. Suppose there is an Adversary $\mathcal{A} = (\widetilde{\mathcal{P}}, \widetilde{\mathcal{V}})$ for which $\mathbf{Adv}(\mathcal{A}, \mathrm{Exp}^{\mathsf{CPLVE}, \lambda, \ell})$ is non-negligible. Thus, it can be used as a subprocedure: either to break security of RSA by taking eth root in $\mathbb{Z}_N^\times$ of a given challenge ciphertext $\widetilde{c}$, or to break GDH for the given instance g, g^α, g^β, by computing $g^{\alpha\beta}$, either with a non-negligible probability. Therefore we draw a bit d which determines our strategy. If $d = 0$, we assume a play against the Adversary in the first scenario, breaking the security of RSA; otherwise, we play against the Adversary in the second scenario, solving the CDH problem.

Init Stage: Let par $\leftarrow \mathbb{G} = (q, g, G)$ and (g, g^α, g^β) be the CDH problem input instance. We set $\mathsf{pk} = g^\alpha$ and give it to $\mathcal{A}$. If $d = 0$, we assume pe and $\widetilde{c}$ to be the RSA input instance, thus we do not know the proper verifier's secret key; otherwise, we honestly generate verifier secret keys $(\mathsf{pe}, \mathsf{se})$. We initiate RO table with columns I, H, r.

Query Stage: We interactively simulate, with an active malicious Verifier $\widetilde{\mathcal{V}}$, the protocol $\pi(\mathcal{P}^{\bar{x}_i}(\mathsf{pk}), \widetilde{\mathcal{V}}^{\mathcal{O}_{\mathcal{H}_G}}(\mathsf{pk}, \bar{x}_i, \{v_{i-1}\}))$, without the secret key, using injected ephemerals $\bar{x}_i$, ℓ times.

Serving Hash queries $\mathcal{O}_{\mathcal{H}_G}(I_i)$**:** If input I_i is in the RO table, the oracle returns the corresponding output H_i. Otherwise, $r_i \leftarrow_R \mathbb{Z}_q^*$, $H_i = g^{r_i}$, add (I_i, H_i, r_i) to the RO table.

(1) **Commitment** $\hat{c}$**:** Receive the commitment $\hat{c}$ in the first message.

(2) **Commitment** X**:** Send $\widetilde{X} = g^{\bar{x}}$ to the Verifier $\widetilde{\mathcal{V}}$.

(3) **Proof** S**:** Upon obtaining m, b from the Verifier, check $m^e \stackrel{?}{=} \hat{c}$ and compute $\bar{c} = \mathcal{H}_q(m, b)$. Query $\mathcal{O}_{\mathcal{H}_G}(\widetilde{X}, \bar{c})$ for r. Set $\widetilde{S} = \widetilde{X}^r \mathsf{pk}^{r\bar{c}} = \hat{g}^{\bar{x}+\mathsf{sk}\bar{c}}$. Note that: $\hat{e}(\widetilde{S}, g) = \hat{e}(\hat{g}, \widetilde{X}\mathsf{pk}^{\bar{c}})$. The *simulated* transcript tuple $\widetilde{T} = (\hat{c}, \widetilde{X}, (m, b), \widetilde{S})$ and the potential *real* protocol execution transcript $T = (\hat{c}, X, (m, b), S)$ are of the same distribution.

Impersonation Stage: The strategy differs between the scenarios:

$d = 0$ We send the challenge ciphertext $\widetilde{c}$ as Verifier's commitment. If the Adversary computes the challenge $c = \mathcal{H}_q(m, b)$ before sending X, we use him to break the security of the underlying encryption scheme. Intercepting query $\mathcal{O}_{\mathcal{H}_q}(m, b)$, we obtain m breaking the *encryption one-wayness*, in this case, being the eth root of $\widetilde{c}$ in $\mathbb{Z}_N^\times$, as $m^e = \widetilde{c}$.

$d = 1$ In ROM, we run $\pi(\widetilde{\mathcal{P}}^{\mathcal{O}_{\mathcal{H}_G}}(\mathsf{pk}, \mathsf{pe}, \{v_i\}), \mathcal{V}(\mathsf{pk}, \mathsf{se}))$ playing the role of the honest Verifier. We use the *rewinding technique*: we fix the random value x used for $X = g^x$ by $\widetilde{\mathcal{P}}$, and upon obtaining a correct proof message, we rewind the prover back to the challenge phase, choosing $b = 0$ in the first run and $b = 1$ in the second run. This gives us $c_1 = \mathcal{H}_q(m, 0)$ for the first run and $c_2 = \mathcal{H}_q(m, 1)$ for the second run. Finally, we get tuples $(\hat{c}, X, m, 0, c_1, S_1, \hat{g}_1, r_1)$ and $(\hat{c}, X, m, 1, c_2, S_2, \hat{g}_2, r_2)$. By inspecting RO tables, we obtain $\hat{g}_1 = \mathcal{O}_{\mathcal{H}_G}(X, c_1) \to g^{\beta r_1}$, $\hat{g}_2 = \mathcal{O}_{\mathcal{H}_G}(X, c_2) \to g^{\beta r_2}$. If we accept the Prover both times, i.e.: $\hat{e}(S_1, g) = \hat{e}(\hat{g}_1, X\mathsf{pk}^{c_1})$ and $\hat{e}(S_2, g) = \hat{e}(\hat{g}_2, X\mathsf{pk}^{c_2})$. Hence we conclude: $S_1 = g^{\beta r_1(x + \alpha c_1)}$ and $S_2 = g^{\beta r_2(x + \alpha c_2)}$. Thus $S_1^{1/r_1}/S_2^{1/r_2} = g^{\beta(\alpha c_1 - \alpha c_2)}$ and $g^{\alpha\beta} = (S_1^{1/r_1}/S_2^{1/r_2})^{1/(c_1 - c_2)}$.

Now, let p denote the non-negligible probability of $\mathcal{A}$ breaking our scheme. Let p_0 be the probability that it knows $c = \mathcal{H}_q(m, b)$ before sending X. Let $p_1 = 1 - p_0$ be the probability that it doesn't know $c = \mathcal{H}_q(m, b)$ before sending X. Thus, we break RSA with probability $\frac{1}{2}pp_0$, or alternatively, we break CDH with probability $\frac{1}{2}p(1 - p_0)$. Hence, we break one of the problems with non negligible probability, which contradicts our assumptions for any probability value $p_0 \in [0, 1]$. □

References

1. Krzywiecki, Ł.: Schnorr-like identification scheme resistant to malicious subliminal setting of ephemeral secret. In: Bica, I., Reyhanitabar, R. (eds.) SECITC 2016. LNCS, vol. 10006, pp. 137–148. Springer, Cham (2016). doi:10.1007/978-3-319-47238-6_10

2. Schnorr, C.P.: Efficient signature generation by smart cards. J. Cryptol. 4(3), 161–174 (1991). http://dx.doi.org/10.1007/BF00196725

3. Okamoto, T.: Provably secure and practical identification schemes and corresponding signature schemes. In: Brickell, E.F. (ed.) CRYPTO 1992. LNCS, vol. 740, pp. 31–53. Springer, Heidelberg (1993). doi:10.1007/3-540-48071-4_3

4. Stinson, D.R., Wu, J.: An efficient and secure two-flow zero-knowledge identification protocol. J. Math. Cryptol. (JMC) 1(3), 201–220 (2007)

5. Wu, J., Stinson, D.R.: An efficient identification protocol and the knowledge-of-exponent assumption. IACR Cryptology ePrint Archive 2007, 479 (2007)

6. Bender, J., Dagdelen, Ö., Fischlin, M., Kügler, D.: The PACE—AA protocol for machine readable travel documents, and its security. In: Keromytis, A.D. (ed.) FC 2012. LNCS, vol. 7397, pp. 344–358. Springer, Heidelberg (2012). doi:10.1007/978-3-642-32946-3_25

7. Fiat, A., Shamir, A.: How to prove yourself: practical solutions to identification and signature problems. In: Odlyzko, A.M. (ed.) CRYPTO 1986. LNCS, vol. 263, pp. 186–194. Springer, Heidelberg (1987). doi:10.1007/3-540-47721-7_12

8. Feige, U., Fiat, A., Shamir, A.: Zero-knowledge proofs of identity. J. Cryptol. 1(2), 77–94 (1988). http://dx.doi.org/10.1007/BF02351717

9. Guillou, L.C., Quisquater, J.-J.: A practical zero-knowledge protocol fitted to security microprocessor minimizing both transmission and memory. In: Barstow, D., et al. (eds.) EUROCRYPT 1988. LNCS, vol. 330, pp. 123–128. Springer, Heidelberg (1988). doi:10.1007/3-540-45961-8_11

10. Kurosawa, K., Heng, S.-H.: Identity-based identification without random oracles. In: Gervasi, O., Gavrilova, M.L., Kumar, V., Laganà, A., Lee, H.P., Mun, Y., Taniar, D., Tan, C.J.K. (eds.) ICCSA 2005. LNCS, vol. 3481, pp. 603–613. Springer, Heidelberg (2005). doi:10.1007/11424826_64

11. Kurosawa, K., Heng, S.-H.: The power of identification schemes. In: Yung, M., Dodis, Y., Kiayias, A., Malkin, T. (eds.) PKC 2006. LNCS, vol. 3958, pp. 364–377. Springer, Heidelberg (2006). doi:10.1007/11745853_24

12. Canetti, R., Goldreich, O., Goldwasser, S., Micali, S.: Resettable zero-knowledge (extended abstract). In: Proceedings of the Thirty-Second Annual ACM Symposium on Theory of Computing (STOC 2000), pp. 235–244 (2000). http://doi.acm.org/10.1145/335305.335334

13. Bellare, M., Fischlin, M., Goldwasser, S., Micali, S.: Identification protocols secure against reset attacks. In: Pfitzmann, B. (ed.) EUROCRYPT 2001. LNCS, vol. 2045, pp. 495–511. Springer, Heidelberg (2001). doi:10.1007/3-540-44987-6_30

14. Ateniese, G., Magri, B., Venturi, D.: Subversion-resilient signature schemes. In: Proceedings of the 22nd ACM SIGSAC Conference on Computer and Communications Security, Denver, 12–6 October 2015, pp. 364–375 (2015)

15. Russell, A., Tang, Q., Yung, M., Zhou, H.: Cliptography: clipping the power of kleptographic attacks. IACR Cryptology ePrint Archive 2015, 695 (2015). http://eprint.iacr.org/2015/695

16. Hanzlik, L., Kluczniak, K., Kutyłowski, M.: Controlled randomness – a defense against backdoors in cryptographic devices. In: Phan, R.C.-W., Yung, M. (eds.) Mycrypt 2016. LNCS, vol. 10311, pp. 215–232. Springer, Cham (2017). doi:10.1007/978-3-319-61273-7_11

17. Raimondo, M.D., Gennaro, R., Krawczyk, H.: Deniable authentication and key exchange. In: Juels, A., Wright, R.N., di Vimercati, S.D.C. (eds.) Proceedings of the 13th ACM Conference on Computer and Communications Security (CCS 2006), Alexandria, 30 October–3 November 2006, pp. 400–409. ACM (2006). http://doi.acm.org/10.1145/1180405.1180454

18. Dwork, C., Naor, M., Sahai, A.: Concurrent zero-knowledge. In: Proceedings of the Thirtieth Annual ACM Symposium on Theory of Computing (STOC 1998), pp. 409–418 (1998). http://doi.acm.org/10.1145/276698.276853

19. Rivest, R.L., Shamir, A., Adleman, L.: A method for obtaining digital signatures and public-key cryptosystems. Commun. ACM 21(2), 120–126 (1978). http://dx.doi.org/10.1145/359340.359342

Evolution of the McEliece Public Key Encryption Scheme

Dominic Bucerzan[1], Vlad Dragoi[2(✉)], and Hervé Talé Kalachi[2,3]

[1] Department of Mathematics and Computer Science,
Aurel Vlaicu University of Arad, 310330 Arad, Romania
dominic@bbcomputer.ro
[2] Laboratoire LITIS - EA 4108 Université de Rouen - UFR Sciences et Techniques,
76800 Saint Etienne du Rouvray, France
{vlad.dragoi1,herve.tale-kalachi1}@univ-rouen.fr
[3] Departament de Mathématiques, Université de Yaounde 1, Yaounde, Cameroun

Abstract. The evolution of the McEliece encryption scheme is a long and thrilling research process. The code families supposed to securely reduce the key size of the original scheme were often cryptanalyzed and thus the future of the code-based cryptography was many times doubted. Yet from this long evolution emerged a great comprehension and understanding of the main difficulties and advantages that coding theory can offer to the field of public key cryptography. Nowadays code-based cryptography has become one of the most promising solutions to post-quantum cryptography. We analyze in this article the evolution of the main encryption variants coming from this field. We stress out the main security issues and point out some new ideas coming from the Rank based cryptography. A summary of the remaining secure variants is given in Fig. 2.

Keywords: Post-quantum cryptography · Coding theory · McEliece encryption scheme

1 Introduction

Code-based cryptography appeared for the first time in 1978, when McEliece proposed the first public key encryption scheme which is not based on number theory primitives [McE78]. Instead he built a scheme for which the security stands on two problems, namely the hardness of the Syndrome Decoding Problem [BMvT78] and the difficulty to distinguish between a binary Goppa code and a random linear code [CFS01,FGO+13]. The scheme disposes of various advantages like

- the complexity of the encryption and decryption algorithms are equivalent to those of symmetric schemes and thus are very efficient compared to other public key schemes.

© Springer International Publishing AG 2017
P. Farshim and E. Simion (Eds.): SecITC 2017, LNCS 10543, pp. 129–149, 2017.
https://doi.org/10.1007/978-3-319-69284-5_10

- the best attacks for solving the Syndrome Decoding Problem are exponential in the code length, which makes code-based schemes of high potential for post-quantum cryptography.

However code-based cryptography came with a big disadvantage: the size of the public keys was about five hundred thousands bits which was unacceptable at that time. Nevertheless the scientific community made a huge progress in reducing the key size of the McEliece PKC by proposing different structures like quasi-cyclic or quasi-dyadic codes. Nowadays the key size is no longer an issue and several practical implementations of the McEliece prove the efficiency and potential of the scheme [BS08, Str10b, CHP12, BCS13, HvMG13, MOG15].

Ever since Peter Shor introduced a polynomial time quantum computer algorithm for factoring integers over $\mathbb{Z}$ and for computing logarithms in the multiplicative group $\mathbb{F}_p$ [Sho94], the code-based cryptography became a serious candidate for public-key cryptography. The interest of the scientific community in this field is nowadays motivated by the latest announcement of the National Institute of Standards and Technology (NIST). They initiated the Post-Quantum crypto Project which aims to define new standards for quantum resistant cryptography and fixed the deadline for public key cryptographic algorithm submissions, for November 2017 (NIST-PQcrypto Project) (http://csrc.nist.gov/groups/ST/post-quantum-crypto/index.html). The purpose of this article is to give a complete evolution of the code-based encryption schemes and rank based encryption schemes. Proposing a global state-of-the-art, that includes both Rank distance and Hamming distance came in a natural manner since there are several facts relating these two topics

- both Hamming distance based schemes and Rank distance schemes sustain their security on the same problem, namely the Syndrome/Rank Syndrome Decoding Problem.
- the similarities do not end here since the properties of the code families that were used are quite equivalent, take for example the case of LRPC (in Rank metric) and LDPC codes (in Hamming metric) or Gabidulin (in Rank metric) and GRS codes (in Hamming metric).
- also the construction techniques are rather similar, for example the QC-LRPC (in Rank metric) and the QC-MDPC (in Hamming metric).

The article also provides a full section dedicated to the security arguments and analyze the main types of attack and it is organized as following. We begin with a preliminary section on the coding theory (Sect. 2). Then we give the necessary details on the McEliece scheme and the actual security arguments for it (see Sect. 3). In Sect. 4 we give the evolution of the McEliece variants starting with the binary Goppa codes up to nowadays. The same analysis is done in Sect. 5, for the Rank based encryption schemes. We conclude with some perspectives in this area.

2   Coding Theory

2.1   Preliminaries

Through this paper, we adopt the following notations: $\mathbb{F}_q$ denotes the finite field with q elements, $\mathsf{GL}_k(\mathbb{F})$ denotes the set of $k \times k$ invertible matrices over a field $\mathbb{F}$. An $[n, k]$ linear code $\mathscr{C}$ over $\mathbb{F}_{q^m}$ is a linear subspace of dimension k of the vector space $\mathbb{F}_{q^m}^n$. Any element in $\mathscr{C}$ is called a *codeword*. A *generator matrix* for a $[n, k]$ linear code is a $k \times n$ matrix (often denoted by $\boldsymbol{G}$) whose rows form a basis for the code. The *dual* of $\mathscr{C}$ denoted by $\mathscr{C}^\perp$ is the linear code which consists of all vectors $\boldsymbol{y} \in \mathbb{F}_{q^m}^n$ such that $\forall\, \boldsymbol{c} \in \mathscr{C} \quad \boldsymbol{y} \cdot \boldsymbol{c}^T = 0$. A *parity-check matrix* of $\mathscr{C}$ is a generator matrix of its dual. It is also a $(n - k) \times n$ matrix $\boldsymbol{H}$ of full rank that satisfies $\boldsymbol{H}\boldsymbol{c}^T = \boldsymbol{0}$ for all $\boldsymbol{c} \in \mathscr{C}$.

Minimum distance of a code. There are several metrics over the vector space $\mathbb{F}_{q^m}^n$ that are known in the literature like the Lee distance, the Hamming distance, the Rank distance etc. In code-based cryptography there are only two of them that became famous: The Hamming distance $\mathsf{d_H}$, that denotes the number of coordinates on which two vectors differ and The Rank distance $\mathsf{d_R}$ defined as follows.

Definition 1 (Rank distance). *The* rank weight *of a vector* $\boldsymbol{x} = (x_1, x_2, ..., x_n)$ *in* $\mathbb{F}_{q^m}^n$ *denoted by* $|\boldsymbol{x}|_q$ *is the dimension of the* $\mathbb{F}_q$*-vector space generated by* $\{x_1, ..., x_n\}$

$$|\boldsymbol{x}|_q = \dim \sum_{i=1}^{n} \mathbb{F}_q x_i.$$

The rank distance $\mathsf{d_R}(\boldsymbol{x}, \boldsymbol{y})$ *is then given by:*

$$\mathsf{d_R}(\boldsymbol{x}, \boldsymbol{y}) = |\boldsymbol{x} - \boldsymbol{y}|_q$$

In the sequel, for a given vector $\boldsymbol{x} \in \mathbb{F}_{q^m}^n$, $|\boldsymbol{x}|$ will denote the Hamming weight of $\boldsymbol{x}$.

Definition 2 (Minimum distance). *The* minimum distance *of a linear code is:*

$$\mathsf{d_{min}}(\mathscr{C}) = \min_{\substack{(\boldsymbol{c}, \boldsymbol{c}^*) \in \mathscr{C} \times \mathscr{C} \\ \boldsymbol{c} \neq \boldsymbol{c}^*}} \mathsf{d}(\boldsymbol{c}, \boldsymbol{c}^*)$$

where d *is any of the aforementioned distances.*

2.2   The General Decoding Problem

The initial purpose of a linear code is to provide an efficient tool for a reliable communication process and it was introduced by Claude Shannon [Sha48]. We explain here a simple case, namely binary linear codes over the Binary Symmetric Channel. Let $\mathscr{C}$ be a $[n, k, d]$ binary linear code with generator matrix $\boldsymbol{G}$ and parity check matrix H, where d is the minimum distance of the code. Encoding a message m into a codeword c is equivalent to compute $c = mG$. Then the codeword c is sent over a BSC(p), where p is the probability of flipping a bit. In other words the receiver obtains $z = c \oplus e \in \mathbb{F}_2^n$ where e is the error vector. The problem the receiver needs to solve here is to recover c from z, which is called the *general decoding problem*.

Since for any codeword c of $\mathscr{C}$ we have $Hc^T = 0_{n-k}$ we deduce that $Hz^T = He^T$. Therefore the dual version of the later problem can be defined generally as follows:

Definition 3 (Syndrome Decoding Problem).
Instance: A full rank matrix $H \in \mathcal{M}_{n-k,n}(\mathbb{F}_{q^m})$, a vector $s \in \mathbb{F}_{q^m}^{n-k}$ and an integer $\omega > 0$.
Question: Is there a vector $x \in \mathbb{F}_{q^m}^n$ of weight $\leq \omega$, such that $Hx^T = s$?

In the case of the Hamming distance we call it the Syndrome Decoding Problem and Rank Syndrome Decoding Problem in the case of the Rank distance. These problems are NP-complete [BMvT78, GZ16].

There are code families for which the later problem is no longer difficult and for which efficient decoding algorithms are known. In the next part we recall some of the linear codes that are used for cryptographic purpose.

2.3 Some Code Families

Reed-Muller codes. The Reed-Muller codes were introduced by David Muller [Mul54] and rediscovered shortly after with an efficient decoding algorithm by Irving Reed [Ree54].[1] The scientific community was highly interested in this family of codes and therefore discovered many structural properties of Reed-Muller codes. Recently Kudekar et al. proved that Reed-Muller codes achieve the capacity of the Erasure channel [KKM+17].

Definition 4 (Reed-Muller codes). *Let m and r be two integers such that $1 \leq r \leq m$ and let $n \stackrel{def}{=} 2^m$. Then the rth order Reed-Muller code $\mathscr{R}(r, m)$ is the binary linear code defined as the set of all vectors $(g(v_1, \ldots, v_m))_{(v_1,\ldots,v_m) \in \mathbb{F}_2^m} \in \mathbb{F}_2^n$, where g ranges over the set of polynomials over $\mathbb{F}_2$ in m variables with degree at most r.*

$$\mathscr{R}(r, m) \stackrel{def}{=} \left\{ (g(v_1, \ldots, v_m))_{(v_1,\ldots,v_m) \in \mathbb{F}_2^m} \mid g \in \mathbb{F}_2[x_1, \ldots, x_m] \ \deg g \leq r \right\}.$$

Generalized Reed-Solomon and Goppa codes. Generalized Reed-Solomon codes, or shortly GRS codes, were introduced by Reed and Solomon in [ISR60] and represent a powerful family of codes with many applications. Ten years after, a new class of codes, binary Goppa codes, was introduced by Valery Goppa [Gop70]. The main reason we detail Goppa codes in the same paragraph with GRS codes is because Goppa codes can be defined as subfield subcodes of GRS codes.

---

[1] Although it seems that these codes were firstly discovered by Mitani in 1951 [Mit51], they became popular only after the article of Muller and Reed.

Definition 5 (Generalized Reed-Solomon codes). *Let k and n be two integers such that $1 \leq k < n \leq q$ where $q = p^m$ is a power of a prime number p. Let $(\boldsymbol{x}, \boldsymbol{y}) \in \mathbb{F}_q^n \times \mathbb{F}_q^n$ be a pair such that $\boldsymbol{x}$ is an n-tuple of distinct elements of $\mathbb{F}_q$ and the elements y_i are nonzero elements in $\mathbb{F}_q$. Then the Generalized Reed-Solomon code $\mathbf{GRS}_k(\boldsymbol{x}, \boldsymbol{y})$ is given by:*

$$\mathbf{GRS}_k(\boldsymbol{x}, \boldsymbol{y}) \stackrel{def}{=} \{(y_1 f(x_1), \ldots, y_n f(x_n)) \mid f \in \mathbb{F}_q[x],\ \deg(f) < k\}.$$

The vector $\boldsymbol{x}$ is called the support of the code and $\boldsymbol{y}$ the multiplier vector. One can easily deduce that a generator matrix of $\mathbf{GRS}_k(\boldsymbol{x}, \boldsymbol{y})$ is given by

$$G = \begin{pmatrix} 1 & 1 & \ldots & 1 \\ x_1 & x_2 & \ldots & x_n \\ \vdots & \vdots & \vdots & \vdots \\ x_1^{k-1} & x_2^{k-1} & \ldots & x_n^{k-1} \end{pmatrix} \begin{pmatrix} y_1 & & & \\ & y_2 & & 0 \\ & & \ddots & \\ 0 & & & y_n \end{pmatrix}.$$

Proposition 1 ([MS86]Theorem 4, Chap. 10). *The dual of a GRS code is also a GRS code and we have*

$$\mathbf{GRS}_k(\boldsymbol{x}, \boldsymbol{y})^{\perp} = \mathbf{GRS}_{n-k}(\boldsymbol{x}, \boldsymbol{z}),$$

where $\boldsymbol{z}$ is a non-zero codeword of the $(n, 1, n)$ GRS code $\mathbf{GRS}_{n-1}(\boldsymbol{x}, \boldsymbol{y})^{\perp}$.

We notice that the vector $\boldsymbol{z}$ with $\forall\ 1 \leq i \leq n, z_i \neq 0$ exists since any non zero codeword of a $[n, 1, n]$ GRS code has a Hamming weight equal to n.

Definition 6 (Alternant codes). *A p-ary alternant code of order r associated to $(\boldsymbol{x}, \boldsymbol{y}) \in \mathbb{F}_{p^m}^n \times \mathbb{F}_{p^m}^n$ denoted by $\mathbf{Alt}_r(\boldsymbol{x}, \boldsymbol{y})$ is*

$$\mathbf{Alt}_r(\boldsymbol{x}, \boldsymbol{y}) \stackrel{def}{=} \mathbf{GRS}_r(\boldsymbol{x}, \boldsymbol{y})^{\perp} \cap \mathbb{F}_p^n.$$

Definition 7 (Binary Goppa codes). *Let $\boldsymbol{x} \in \mathbb{F}_{2^m}^n$ be a $n - tuple$ of distinct elements and $g \in \mathbb{F}_{2^m}[x]$ be a polynomial of degree t such that $\forall\ i, g(x_i) \neq 0$. Let $\boldsymbol{y} \stackrel{def}{=} (1/g(x_i), \ldots, 1/g(x_n))$ then the binary Goppa code is defined by*

$$\Gamma(\boldsymbol{x}, g) \stackrel{def}{=} \mathbf{Alt}_t(\boldsymbol{x}, \boldsymbol{y}).$$

There are several decoding techniques for Goppa codes like for example the Berlekamp-Massey algorithm, the Extended Euclidean Algorithm or the Patterson algorithm [MS86, Chap. 12].

LDPC and MDPC codes. Another important class of linear codes is the family of low density parity check (LDPC) codes discovered by Gallager [Gal63]. He was motivated by the problem of finding "random-like" codes that could be decoded near the channel capacity with quasi-optimal performance and feasible complexity. Since LDPC were too complex for the technology at that time, they were forgotten for more than 30 years, and rediscovered by MacKay [Mac99] and Sipser and Spielman [SS96]. These codes were extended in a natural way to moderate density parity check codes in [OB09]. LDPC codes have many applications in communication field as well as in cryptography.

Definition 8 (LDPC/MDPC codes). *A (n, k, ω)-code is a linear code defined by a $k \times n$ parity-check matrix $(k < n)$ where each row has weight ω.*

A LDPC code is a (n, k, ω)-code with $\omega = O(1)$, when $n \to \infty$. [Gal63]
A MDPC code is a (n, k, ω)-code with $\omega = O(\sqrt{n})$, when $n \to \infty$. [OB09]

The theory of error correcting codes is not only a highly important tool in the communication field, it is also applied to public key cryptography. One of the oldest public key encryption scheme, namely the McEliece PKC [McE78], is based on several aspects from coding theory.

3 McEliece and Niederreiter Encryption Scheme

3.1 Description

The McEliece public key encryption scheme [McE78] is composed of three algorithms: *key generation* (KeyGen), *encryption* (Encrypt) and *decryption* (Decrypt). The key generation algorithm takes as input the integers n, m, k, t, q such that $k < n$ and $t < n$ and outputs the public key/private key pair (pk, sk).

KeyGen$(n, m, k, t, q) = (\text{pk}, \text{sk})$

1. Pick a generator matrix $\boldsymbol{G}$ of a $[n, k]$ code $\mathscr{C}$ that can corrects t errors.
2. Pick at random $\boldsymbol{S}$ in $\mathrm{GL}_k(\mathbb{F}_{q^m})$ and a $n \times n$ permutation matrix $\boldsymbol{P}$.
3. Compute $\boldsymbol{G}_{\text{pub}} \overset{\text{def}}{=} \boldsymbol{SGP}$.
4. Return

$$\text{pk} = (\boldsymbol{G}_{\text{pub}}, t) \text{ and } \text{sk} = (\boldsymbol{S}, \boldsymbol{P}).$$

In order to encrypt a message $\boldsymbol{m} \in \mathbb{F}_{q^m}^k$ one applies the following function

Encrypt$(\boldsymbol{m}, \text{pk}) = \boldsymbol{z}$

1. Generate a random error-vector $\boldsymbol{e} \in \mathbb{F}_{q^m}^n$ with $|\boldsymbol{e}| \leq t$
2. Return $\boldsymbol{z} = \boldsymbol{m}\boldsymbol{G}_{\text{pub}} \oplus \boldsymbol{e}$

The decryption takes as input a ciphertext $\boldsymbol{z}$ and the private key sk and outputs the corresponding message $\boldsymbol{m}$

Decrypt$(\boldsymbol{z}, \text{sk}) = \boldsymbol{m}$

1. Compute $\boldsymbol{z}^* = \boldsymbol{z}\boldsymbol{P}^{-1}$ and $\boldsymbol{m}^* = Decode(\boldsymbol{z}^*, \boldsymbol{H})$
2. Return $\boldsymbol{m}^*\boldsymbol{S}^{-1}$.

$Decode(.,.)$ is an efficient decoding algorithm for $\mathscr{C}$. Notice that multiplying the error vector by a permutation does not change the weight of the vector. One can easily verify the correctness of the scheme by checking

$$\text{Decrypt}(\text{Encrypt}(\boldsymbol{m}, \text{pk}), \text{sk}) = \boldsymbol{m}.$$

The Niederreiter public-key encryption scheme [Nie86] is similar to the McEliece's scheme. It uses the dual code and thus the public key is a parity check matrix for the code. The message will be an error vector that is encrypted into a syndrome. In [LDW94] it is showed that the two schemes are equivalent in term of security.

3.2   Security Arguments

The security of all the variants à la McEliece is based on two facts: firstly the public code is supposed to be indistinguishable from a random code. If the later supposition is satisfied then in order to decrypt a cyphertext one has to solve the Syndrome Decoding Problem for a random code (see Definition 3), which is known as a difficult problem. There are three types of attacks known in the literature: Distinguishing Attacks, Message Recovery Attacks (MRA) and Key Recovery Attacks (KRA).

Distinguishing attacks. Even though the indistinguishably of the public code in the original McEliece scheme was not proved, there is a strong believe that this problem is hard. However, a recent breakthrough in this area was the distinguisher for high rate Goppa codes, proposed in [FGO+13]. It is based on the star product of two codes and uses the dimension of the square code in order to distinguish between a random linear code and a high rate Goppa code. This technique also works on high rate Alternant codes [FGO+13], Reed-Solomon codes [CGG+14], Reed-Muller codes [CB13,OTK15] etc.

Message Recovery Attacks. In this scenario an adversary aim to recover the plaintext from a given ciphertext. If the public code is indistinguishable from a random code then the MRA become equivalent to solving the Syndrome Decoding Problem. The most efficient algorithm to solve the Syndrome Decoding Problem is the Information Set Decoding (ISD). Details about the different variants of ISD and their complexity analysis are given in [CTS16]. However, the best variant has a complexity which is exponential in the codes parameters.

Key Recovery Attacks. The key recovery adversary aims to retrieve the private key from a given public key. If the cryptanalyst manages to efficiently recover the private key, then he can also decode and find all the messages that have been encrypted with that key. Therefore it is considered as the most powerful possible attack. In the KRA scenario the adversary is often reduced to solve the following problem.

Definition 9 (Permutation Code Equivalence Problem). *Let G and G^* be the generating matrices for two $[n, k]$ binary linear codes. Given G and G^* does there exist a $k \times k$ binary invertible matrix S and $n \times n$ permutation matrix P such that $G^* = SGP$?*

The computational problem was studied by Petrank and Roth over the binary field [PR97], in which the authors proved that the problem is not NP-complete. The most common algorithm used to solve this problem is the Support Splitting Algorithm (SSA) [Sen00]. This algorithm is very efficient in the random case, but cannot be used in the case of codes with large Hulls or codes with large Permutation group such as Goppa codes, Reed-Muller codes, ... When the SSA is infeasible, other efficient technique can be employed such as the *Minimum Weight Codewords* approach. The idea is to use the subcode spanned by the set of minimum weight codewords and solve the code equivalence problem for the later code. Indeed, in the case of many linear codes, the code spanned by the set of minimum weight codewords is almost the entire code. This is the case of Polar codes and more generally of any Decreasing Monomial codes (see [BDOT16]). This technique was used to solve the code equivalence problem for Reed-Muller codes [MS07] and Polar codes [BCD+16]. The main step of this technique is the minimum weight codewords searching. The most efficient algorithms for this are derived from the Information Set Decoding algorithm.

Side-channel attacks. The importance of practical issues is crucial for designing a cryptosystem. A designer should be able to prove that the scheme can be securely implemented and that eventual side-channel attacks can easily be countered. In this scenario the attacker has the capability to access and monitor different parameters of the implementation, like for example a particular function in the decryption process. In a successful side-channel attack, the aforementioned advantage reveals information on the private message or on the private key of the scheme.

4   McEliece Variants

In the previous section, several security issues are revealed, fact that raised a fundamental question: *What is the most appropriate code family for the McEliece scheme?*

4.1   Binary Irreducible Goppa Codes

They were proposed in the original paper of McEliece [McE78]. Even though the original parameters were broken in [BLP08], they proposed a new set of secured parameters (see Fig. 1). Despite their well known structure there are no efficient key recovery or decoding attacks against binary irreducible Goppa codes. A **distinguisher** exists in the case of high rate Goppa codes [FGO+13]. But despite of this potential vulnerability there is no efficient algorithm for the moment exploiting the knowledge and the properties of the distinguisher. The existence of **weak keys** for Goppa codes was raised by Sendrier and Loidreau in [LS01].

We notice from Fig. 1 that the size of the public key is a real disadvantage of the McEliece scheme compared to the well known RSA encryption scheme [RSA78]. Therefore reducing the size of the keys is one of the starting points of a continuous research interest in this field. We mention the existence of a

Security level(-bit)	$[n,k]$	t	Public Key size (bits)	RSA - Public key size (bits)
80	[1632, 1269]	33	460647	512
128	[2960, 2288]	56	1537536	3072
256	[6624, 5129]	115	7667855	15360

Fig. 1. Parameters and key size for McEliece with Goppa codes from [BLP08] and key size for the RSA scheme

compact variant of the McEliece scheme based on quasi-dyadic Goppa codes due to Misoczki and Barreto [MB09], variant that is not yet broken in the binary case. The binary Goppa codes were also the most cryptanalyzed scheme from side-channel perspective. There are mainly two types of side-channel attacks classified by their goal:

1. Recover the secret message [STM+08, AHPT11];
2. Recover the private key (fully or partially) [Str13, Str10a, SSMS09, BCDR16].

In each article the authors propose to counter the leak and thus step towards a secure implementation of the scheme. Countermeasures and secured implementations are also proposed in [CHP12, DCCR13, BCS13].

4.2   Generalized Reed-Solomon Codes

This family was proposed for the first time by Niederreiter in [Nie86] but turned out to be an insecure solution. Indeed, six years after the article was published, Sidelnikov and Shestakov proposed a polynomial time attack against this variant [SS92]. Nevertheless the idea of using GRS codes was reconsidered more than ten years after by Berger and Loidreau when they proposed to consider subcodes of GRS codes [BL05]. Unfortunately this technique was also attacked in two steps by Wieschebrink [Wie06a, Wie09], using the **square code structure**.

Other attempts to repair the Niederreiter variant were proposed by Wieschebrink [Wie06b] who's idea was to add random column to the generator matrix. But this variant turned out to be extremely unsecure against **square code type attacks** [CGG+14]. Nevertheless GRS codes are still of high interest for this community since several modified version of the McEliece scheme use this family of codes. For example Baldi et al. [BBC+16] proposed to change the permutation matrix, Tillich et al. [MCT16] propose to use them in a *"u | u + v"* construction, Wang [Wan16] propose to use a technique derived from Wieschebrink's idea.

4.3   Reed-Muller Codes

Reed-Muller codes were proposed by Sidelnikov's in [Sid94] and was firstly attacked by Minder and Shokrollahi [MS07]. In the case of Reed-Muller codes the Key Recovery Attack is reduced to solving the code equivalence problem since there is only one $\mathscr{R}(r, m)$. Minder and Shokrollahi managed to solve this problem using a filtration type attack based on the structure properties of the minimum weight codewords. The complexity of their algorithm was dominated by the minimum weight codewords searching algorithm.

Recently, Chizhov and Borodin [CB14] proposed another attack that could solve the code equivalence problem, for some of the parameters of the Reed-Muller codes, in polynomial time. Their idea was to use two simple operations in order to find the first order Reed-Muller code given the rth order Reed-Muller code. Indeed they noticed that the dual and the **square code** of a Reed-Muller code is still a Reed-Muller code. So they combined these operations in order to approach the $\mathscr{R}(1, m)$. A modified version using the masking technique introduced by Wieschebrink was proposed in [GM13] and recently broken by Otmani and Talé-Kalachi [OTK15] using a **square code** type attack.

4.4 Algebraic-Geometry Codes

This family of codes was suggested by Janwa and Moreno [JM96]. Several articles discussed the potential vulnerabilities of this variant and proposed algorithms that could be deployed to attack in some particular cases [FM08, SS92]. Nevertheless they can not be generalized and suffer in terms of efficiency. In [CMCP14] Couvreur, Marquez-Corbella and Pellikaan proposed a polynomial type algorithm that works on codes from curves of arbitrary genus.

4.5 Concatenated Codes

Concatenated codes were the first family of probabilistic codes analyzed from a cryptographic point of view. Sendrier detailed in [Sen94, Sen98] the main vulnerabilities of ordinary concatenated codes.

4.6 LDPC Codes

Monico, Rosenthal and Shokrollahi were the first ones to propose and analyze a McEliece variant using low density parity check codes in [MRAS00]. Using the idea of Gaborit to consider quasi-cyclic codes [Gab05][2] the new QC-LDPC cryptosystem was presented by Baldi and Chiaraluce in [BC07]. Both BCH codes and LDPC codes with quasi-cyclic structure were successfully cryptanalyzed by Otmani, Tillich and Dallot [OTD08]. In order to prevent the last attack, a modification based on increasing the weight of the codewords in the public code was proposed in [BBC08]. In the book of Baldi [Bal14] all the details about the thrilling combats defeating and attacking the LDPC codes are given.

4.7 Wild Goppa Codes

This code family is a natural extension from binary Goppa codes to non binary fields. It was proposed by Bernstein, Lange and Peters in [BLP10] and [BLP11]. Many of the proposed parameters were broken by Couvreur, Otmani and Tillich using **filtration type techniques** [COT14a, COT14b], for quadratic extensions.

4.8 Srivastava Codes

Srivastava codes were proposed in [Per12] in order to reduce the key length of the original McEliece scheme. The author uses Quasi-Dyadic Srivastava codes and gives another application of these types of codes for signature schemes. Even though the parameters for the signature were broken in [FOP+16], the parameters for the encryption scheme are still valid.

---

[2] In [Gab05] the author proposes BCH codes with quasi-cyclic structure. The idea of adding the quasi cyclic structure became one of the main techniques for reducing the key size in the McEliece scheme.

4.9 MDPC Codes

Moderate Density Parity-Check codes are probably the most suitable codes in a McEliece type scheme [MTSB13]. Many cryptographic arguments are in favour of this family of codes like efficiency, small key size when used with a quasi-cyclic structure and the most important to our opinion the lack of algebraic structure. Another security argument is the fact that the usual distinguisher does not work for MDPC codes. In a recent paper, weak keys of the QC-MDPC scheme are revealed [BDLO16]. However the authors show how to avoid vulnerable parameters.

4.10   Convolutional Codes

Convolutional codes represented among the shortest term solutions since between the proposed article by Londahl and Johansson [LJ12] and the efficient attack by Landais and Tillich [LT13] only one year passed.

4.11   Polar Codes

This family of codes was as unfortunate as convolutional code. The first variant using Polar codes was proposed by Shrestha and Kim [SK14] while the second one using subcodes of Polar codes was given in [HSEA14]. In [BCD+16] the first variant was attacked using the structure of the minimum weight codewords. The authors managed to solve the code equivalence problem for Polar codes and thus completely break the scheme.

To close this section we emphasize that there are code families which are not appropriate in this context due to their structural properties, namely the GRS codes, the Reed-Muller codes, the Polar codes ... However several classes of codes remain secure in a McEliece PKC such as original binary Goppa codes, LDPC and MDPC codes etc. A complete summary of the remaining secure code families is also given in Fig. 2. Meanwhile the scientific community developed a new idea, that consists in working with another metric, for instance the rank-metric. Nowadays, this part of the public-key cryptography is known under the name of rank-based cryptography.

5   Rank Based Encryption Schemes

The first rank-metric scheme was proposed in [GPT91] by Gabidulin, Paramonov and Tretjakov which is now called the GPT cryptosystem. This scheme can be seen as an analogue of the McEliece public key cryptosystem based on the class of Gabidulin codes. In the following, we present the class Gabidulin codes. In order to simplify the notation, for any x in $\mathbb{F}_{q^m}$ and for any integer i, the quantity x^{q^i} is denoted by $x^{[i]}$.

Definition 10 (Gabidulin code). *Let $g \in \mathbb{F}_{q^m}^n$ such that $|g|_q = n$. The $(n, k)-$ Gabidulin code denoted by $\mathscr{G}_k(g)$ is the code with a generator matrix G where:*

$$G = \begin{pmatrix} g_1^{[0]} & \cdots & g_n^{[0]} \\ \vdots & & \vdots \\ g_1^{[k-1]} & \cdots & g_n^{[k-1]} \end{pmatrix}. \tag{1}$$

A matrix of the form (1) is called a $q-$ Vandermonde matrix. Gabidulin codes are known to have a good decoding capability [GPT91].

5.1   The General GPT Cryptosystem

The key generation algorithm of the general GPT cryptosystem takes as input the integers k, ℓ, n and m such that $k < n \leq m$ and $\ell \ll n$ and outputs the public key/private key pair $(\mathsf{pk}, \mathsf{sk})$.

$\mathsf{KeyGen}(n, m, k, \ell, q) = (\mathsf{pk}, \mathsf{sk})$

1. Let $G \in \mathcal{M}_{k,n}(\mathbb{F}_{q^m})$ be a generator matrix of the Gabidulin code $\mathscr{G}_k(g)$
2. Pick $S \in \mathsf{GL}_k(\mathbb{F}_{q^m})$, $X \in \mathcal{M}_{k,\ell}(\mathbb{F}_{q^m})$ and $P \in \mathsf{GL}_{n+\ell}(\mathbb{F}_q)$.
3. Compute $G_{\text{pub}} \overset{\text{def}}{=} S(X \mid G)P$ and $t = \frac{n-k}{2}$
4. Return

$$\mathsf{pk} = (G_{\text{pub}}, t) \text{ and } \mathsf{sk} = (S, P).$$

To encrypt a message $m \in \mathbb{F}_{q^m}^k$, apply the following function

$\mathsf{Encrypt}(m, \mathsf{pk}) = z$

1. Generate a random error-vector $e \in \mathbb{F}_{q^m}^n$ with $|e|_q \leq t$
2. Return $z = mG_{\text{pub}} \oplus e$

The decryption takes as input a ciphertext z and the private key sk and outputs the corresponding message m. $\mathsf{Decrypt}(z, \mathsf{sk})$ firstly computes $zP^{-1} = mS(X \mid G) + eP^{-1}$. The last n components of zP^{-1} will satisfy $z' = mSG + e'$ where e' is a sub-vector of eP^{-1} hence $|e'|_q \leq t$. It then applies a fast decoding algorithm of $\mathscr{G}_k(g)$ to z' and obtain mS and hence m.

Security. In [Ove08], Overbeck proposed a very efficient attack on the GPT cryptosystem. Several works propose to resist to Overbeck's attack either by taking a column scrambler matrix defined over the extension field $\mathbb{F}_{q^m}$ [Gab08, GRH09, RGH11, GP14] or by taking special distortion matrix as in [Loi10, RGH10]. We describe in the following all existing variant of the GPT cryptosystem after the apparition of Overbeck's attacks, and we give the state of the security of each variant.

5.2 GPT Cryptosystem with Column Scrambler on the Extension Field

The first paper that consider column scrambler matrix over the extension field is Gabidulin's paper [Gab08]. The important points are Key generation and decryption; the encryption phase is without change. The author proposed to describe the system as follows:

Description of the Scheme. The key generation algorithm works as for the general GPT scheme, with the difference: P in $\mathsf{GL}_{n+\ell}(\mathbb{F}_{q^m})$ is such that there exist Q_{11} in $\mathcal{M}_{\ell,\ell}(\mathbb{F}_{q^m})$, Q_{21} in $\mathcal{M}_{n,\ell}(\mathbb{F}_{q^m})$, Q_{22} in $\mathcal{M}_{n,n}(\mathbb{F}_q)$ and Q_{12} in $\mathcal{M}_{\ell,n}(\mathbb{F}_{q^m})$ with $|Q_{12}| = s < t$ so that

$$P^{-1} = \left(\begin{bmatrix} Q_{11} & Q_{12} \\ Q_{21} & Q_{22} \end{bmatrix} \right). \tag{2}$$

The public key is $(G_{\mathrm{pub}}, t_{\mathrm{pub}})$ with $t_{\mathrm{pub}} = t - s$ and $G_{\mathrm{pub}} = S\,(X \mid G)\,P$.

Decryption. We have $cP^{-1} = mS\,(X \mid G) + eP^{-1}$. Suppose that $e = (e_1 \mid e_2)$ where $e_1 \in \mathbb{F}_{q^m}^{\ell}$ and $e_2 \in \mathbb{F}_{q^m}^{n}$. We have:

$$eP^{-1} = (e_1 Q_{11} + e_2 Q_{21} \mid e_1 Q_{12} + e_2 Q_{22}) \tag{3}$$

It is clear that

$$|e_1 Q_{12} + e_2 Q_{22}| \le |e_1 Q_{12}| + |e_2 Q_{22}| \le s + t - s.$$

So the plaintext m is recovered by applying the decoding algorithm only to the last n components of cP^{-1}.

Several authors also proposed other constructions of the column scrambler on the extension field. In [GRH09, RGH11] it is proposed for instance to choose a column scrambler matrix $P^* = TP$ such that

$$P^{-1} = (Q_1 \mid Q_2) \tag{4}$$

where $Q_1 \in \mathcal{M}_{n,s}(\mathbb{F}_{q^m})$ while $Q_2 \in \mathcal{M}_{n,(n-s)}(\mathbb{F}_q)$. This construction can be seen as a variant of the more general construction given in [Gab08] (see [OTKN16] for more details). In [GP13, GP14], another variant is also proposed. This variant consists to use a column scrambler matrix P such that

$$P^{-1} = T + Z \tag{5}$$

$T \in \mathsf{GL}_{n+\ell}(\mathbb{F}_q)$ and $Z \in \mathcal{M}_{n+\ell,n+\ell}(\mathbb{F}_{q^m})$ with $|Z| = s$. However, this last variant was shown in [UG14] to be equivalent to the general GPT cryptosystem [GO01] and hence not secure.

Security. In was recently shown in [OTKN16] that the Gabidulin's general construction [Gab08] is not secured, even if a more general column scrambler $P^* = TPQ$ is considered ($T, Q \in \mathsf{GL}_{n+\ell}(\mathbb{F}_q)$ and P being a matrix that the inverse is given by Eq. 2). This attack also implies and attack on the variant of [GRH09, RGH11] since the construction of [Gab08] is a generalization of the constructions given in [GRH09, RGH11, GP14, GP13].

5.3 GPT Cryptosystems with a Special Distortion Matrix

Loidreau reparation. The main objective of the Loidreau reparation [Loi10] is not to propose a new system, but to propose parameters that would prevent Overbeck's attack. The idea is to take a very large ℓ ($\ell >>> n - k$) and use a matrix $X \in \mathcal{M}_{k,\ell}(\mathbb{F}_{q^m})$ with a very low rank s such that $s(n-k) \leq \ell - a$ where a is a given integer. Even if the keys sizes of this reparation are small compared to what we have in the McEliece encryption scheme [McE78], they remain very large. It is the reason why the author of [RGH10] proposed the "smart approach" that aim to avoid Overbeck's attack while keeping small keys sizes.

The smart approach. As in the Loidreau's reparation, the only difference is on the generation of X. The authors proposed to take a distortion matrix $X \in \mathcal{M}_{k,\ell}(\mathbb{F}_{q^m})$ that is a concatenation of a $q-$Vandermonde matrix $X_1 \in \mathcal{M}_{k,a}(\mathbb{F}_{q^m})$ and a random matrix $X_2 \in \mathcal{M}_{k,\ell-a}(\mathbb{F}_{q^m})$ with $0 < a < \ell$. More precisely, to design the public generator matrix, let $S \in \mathsf{GL}_k(\mathbb{F}_{q^m})$, $X_2 \in \mathcal{M}_{k,\ell-a}(\mathbb{F}_{q^m})$, $b = (b_1, \cdots, b_a)$ and

$$X_1 = \begin{pmatrix} b_1^{[0]} & \cdots & b_a^{[0]} \\ \vdots & & \vdots \\ b_1^{[k-1]} & \cdots & b_a^{[k-1]} \end{pmatrix}. \tag{6}$$

Select $P \in \mathsf{GL}_{n+\ell}(\mathbb{F}_q)$ and compute

$$G_{\text{pub}} = S\,(X_1 \mid X_2 \mid G)\,P$$

Security. A successful cryptanalysis of the previous variants was recently propose in [HMR15]. We also emphasise that there is a recent Message Recovery Attack against the aforementioned variants by [GRS16, HTMR16].

5.4 LRPC Cryptosystem

Beside the Gabidulin codes and inspired by the class of MDPC/LDPC codes in Hamming metric, a new class of rank metric codes was recently proposed in [GMRZ13] namely Low Rank Parity Check codes. They are the adaptation of the MDPC/LDPC codes in the rank metric. The LRPC cryptosystem [GMRZ13] is thus the analogue of the MDPC McEliece scheme. The main advantage of the scheme is that it comes, as the MDPC PKC, with a quasi-cyclic version, which allows to drastically reduce the key size. The QC-LRPC scheme is therefore one of the most promising rank-based encryption scheme since it has many security arguments in its favour: compared to the Gabidulin codes, the LRPC codes have a weak algebraic structure and thus seem much more fitted for a cryptographic purpose. Secondly the QC-LRPC scheme is equivalent to the NTRU [HPS98] and thus benefit of a quite long research experience from a cryptanalytic point of view.

Security level(-bit)	Binary Goppa [BLP08]	Wild Goppa [BLP10]	QD - Srivastava [Per12]	QC - LDPC [Bal14]	QC - MDPC [MTSB13]	LRPC [GMRZ13]
80	460647	-	36288	-	4801	1681
128	1537536	1523278	37440	12351	9857	2809

Fig. 2. Key size in bits for the remaining secure code families in the McEliece scheme

6   Conclusion and Perspectives

In this article we have given a state-of-the-art of the McEliece encryption scheme. We have also detailed the main security threats for the scheme and for each of the mentioned variants. The general idea is to choose an appropriate private code that will be masked into a public one. This technique opens a general security question of indistinguishability of the public code from a random code. Even though several variants remain secured against existing attacks there is no theoretical guaranty of their security. By that we mean there is no security proof for the aforementioned variants. For instance there is no formal proof of the indistinguishability of the public code from a random one. The table bellow summarizes the remaining secure code families in the McEliece scheme. We emphasize that this table is not complete, but the variants given are the principal ones known with parameters.

Following McEliece's idea a possible solution for this problem would be to find a new masking technique for which there is a formal proof of the indistinguishability of the public code from a random one. In [Wan16] the author propose a masking technique for which he proves that the public code is equivalent to a random code and thus reintroduce in the context all the structural codes that have been broken. Another solution was already proposed by Alekhnovich who proposed an innovative approach based on the difficulty of decoding purely random codes [Ale11]. Several authors were inspired by his work [DMN12, DV13, KMP14, ABD+16]. This two approaches open a new perspective for code-based cryptography.

References

[ABD+16] Aguilar, C., Blazy, O., Deneuville, J.-C., Gaborit, P., Zémor, G.: Efficient encryption from random quasi-cyclic codes. arXiv preprint (2016). arXiv:1612.05572

[AHPT11] Avanzi, R., Hoerder, S., Page, D., Tunstall, M.: Side-channel attacks on the McEliece and Niederreiter public-key cryptosystems. J. Cryptogr. Eng. **1**(4), 271–281 (2011)

[Ale11] Alekhnovich, M.: More on average case vs approximation complexity. Comput. Complex. **20**(4), 755–786 (2011)

[Bal14] Baldi, M.: QC-LDPC Code-Based Cryptography. SpringerBriefs in Electrical and Computer Engineering, p. 120. Springer, Heidelberg (2014). doi:10.1007/978-3-319-02556-8

[BBC08]   Baldi, M., Bodrato, M., Chiaraluce, F.: A new analysis of the mceliece cryptosystem based on QC-LDPC codes. In: Ostrovsky, R., De Prisco, R., Visconti, I. (eds.) SCN 2008. LNCS, vol. 5229, pp. 246–262. Springer, Heidelberg (2008). doi:10.1007/978-3-540-85855-3_17

[BBC+16] Baldi, M., Bianchi, M., Chiaraluce, F., Rosenthal, J., Schipani, D.: Enhanced public key security for the mceliece cryptosystem. J. Cryptol. **29**(1), 1–27 (2016)

[BC07]   Baldi, M., Chiaraluce, F.: Cryptanalysis of a new instance of McEliece cryptosystem based on QC-LDPC codes. In: Proceedings of IEEE International Symposium on Information Theory - ISIT, pp. 2591–2595, Nice, France, June 2007

[BCD+16] Bardet, M., Chaulet, J., Dragoi, V., Otmani, A., Tillich, J.-P.: Cryptanalysis of the McEliece public key cryptosystem based on polar codes. In: Takagi, T. (ed.) PQCrypto 2016. LNCS, vol. 9606, pp. 118–143. Springer, Cham (2016). doi:10.1007/978-3-319-29360-8_9

[BCDR16] Bucerzan, D., Cayrel, P.-L., Dragoi, V., Richmond, T.: Improved timing attacks against the secret permutation in the mceliece PKC. Int. J. Comput. Commun. Control **12**(1), 7–25 (2016)

[BCS13]   Bernstein, D.J., Chou, T., Schwabe, P.: McBits: fast constant-time code-based cryptography. In: Bertoni, G., Coron, J.-S. (eds.) CHES 2013. LNCS, vol. 8086, pp. 250–272. Springer, Heidelberg (2013). doi:10.1007/978-3-642-40349-1_15

[BDLO16] Bardet, M., Dragoi, V., Luque, J.-G., Otmani, A.: Weak keys for the quasi-cyclic MDPC public key encryption scheme. In: Pointcheval, D., Nitaj, A., Rachidi, T. (eds.) AFRICACRYPT 2016. LNCS, vol. 9646, pp. 346–367. Springer, Cham (2016). doi:10.1007/978-3-319-31517-1_18

[BDOT16] Bardet, M., Dragoi, V., Otmani, A., Tillich, J.-P.: Algebraic properties of polar codes from a new polynomial formalism. In: IEEE International Symposium on Information Theory (ISIT 2016), Barcelona, Spain, 10–15 July 2016, pp. 230–234 (2016)

[BL05]   Berger, T.P., Loidreau, P.: How to mask the structure of codes for a cryptographic use. Des. Codes Cryptogr. **35**(1), 63–79 (2005)

[BLP08]   Bernstein, D.J., Lange, T., Peters, C.: Attacking and defending the McEliece cryptosystem. In: Buchmann, J., Ding, J. (eds.) PQCrypto 2008. LNCS, vol. 5299, pp. 31–46. Springer, Heidelberg (2008). doi:10.1007/978-3-540-88403-3_3

[BLP10]   Bernstein, D.J., Lange, T., Peters, C.: Wild McEliece. In: Biryukov, A., Gong, G., Stinson, D.R. (eds.) SAC 2010. LNCS, vol. 6544, pp. 143–158. Springer, Heidelberg (2011). doi:10.1007/978-3-642-19574-7_10

[BLP11]   Bernstein, D.J., Lange, T., Peters, C.: Wild McEliece incognito. In: Yang, B.-Y. (ed.) PQCrypto 2011. LNCS, vol. 7071, pp. 244–254. Springer, Heidelberg (2011). doi:10.1007/978-3-642-25405-5_16

[BMvT78] Berlekamp, E., McEliece, R., van Tilborg, H.: On the inherent intractability of certain coding problems. IEEE Trans. Inform. Theory **24**(3), 384–386 (1978)

[BS08]   Biswas, B., Sendrier, N.: McEliece cryptosystem implementation: theory and practice. In: Buchmann, J., Ding, J. (eds.) PQCrypto 2008. LNCS, vol. 5299, pp. 47–62. Springer, Heidelberg (2008). doi:10.1007/978-3-540-88403-3_4

[CB13] Chizhov, I.V., Borodin, M.A.: The failure of McEliece PKC based on Reed-Muller codes. IACR Cryptology ePrint Archive, Report 2013/287 (2013). http://eprint.iacr.org/

[CB14] Chizhov, I.V., Borodin, M.A.: Effective attack on the McEliece cryptosystem based on Reed-Muller codes. Discr. Math. Appl. 24(5), 273–280 (2014)

[CFS01] Courtois, N.T., Finiasz, M., Sendrier, N.: How to achieve a McEliece-based digital signature scheme. In: Boyd, C. (ed.) ASIACRYPT 2001. LNCS, vol. 2248, pp. 157–174. Springer, Heidelberg (2001). doi:10.1007/3-540-45682-1_10

[CGG+14] Couvreur, A., Gaborit, P., Gauthier-Umaña, V., Otmani, A., Tillich, J.-P.: Distinguisher-based attacks on public-key cryptosystems using Reed-Solomon codes. Des. Codes Cryptogr. 73(2), 641–666 (2014)

[CHP12] Cayrel, P.-L., Hoffmann, G., Persichetti, E.: Efficient implementation of a CCA2-secure variant of mceliece using generalized srivastava codes. In: Fischlin, M., Buchmann, J., Manulis, M. (eds.) PKC 2012. LNCS, vol. 7293, pp. 138–155. Springer, Heidelberg (2012). doi:10.1007/978-3-642-30057-8_9

[CMCP14] Couvreur, A., Márquez-Corbella, I., Pellikaan, R.: A polynomial time attack against algebraic geometry code based public key cryptosystems. In: Proceedings of IEEE International Symposium on Information Theory (ISIT 2014), pp. 1446–1450, June 2014

[COT14a] Couvreur, A., Otmani, A., Tillich, J.-P.: New identities relating wild Goppa codes. Finite Fields Appl. 29, 178–197 (2014)

[COT14b] Couvreur, A., Otmani, A., Tillich, J.-P.: Polynomial time attack on wild McEliece over quadratic extensions. In: Nguyen, P.Q., Oswald, E. (eds.) EUROCRYPT 2014. LNCS, vol. 8441, pp. 17–39. Springer, Heidelberg (2014). doi:10.1007/978-3-642-55220-5_2

[CTS16] Canto Torres, R., Sendrier, N.: Analysis of information set decoding for a sub-linear error weight. In: Takagi, T. (ed.) PQCrypto 2016. LNCS, vol. 9606, pp. 144–161. Springer, Cham (2016). doi:10.1007/978-3-319-29360-8_10

[DCCR13] Dragoi, V., Cayrel, P.-L., Colombier, B., Richmond, T.: Polynomial structures in code-based cryptography. In: Paul, G., Vaudenay, S. (eds.) INDOCRYPT 2013. LNCS, vol. 8250, pp. 286–296. Springer, Cham (2013). doi:10.1007/978-3-319-03515-4_19

[DMN12] Döttling, N., Müller-Quade, J., Nascimento, A.C.A.: IND-CCA secure cryptography based on a variant of the LPN problem. In: Wang, X., Sako, K. (eds.) ASIACRYPT 2012. LNCS, vol. 7658, pp. 485–503. Springer, Heidelberg (2012). doi:10.1007/978-3-642-34961-4_30

[DV13] Duc, A., Vaudenay, S.: HELEN: a public-key cryptosystem based on the LPN and the decisional minimal distance problems. In: Youssef, A., Nitaj, A., Hassanien, A.E. (eds.) AFRICACRYPT 2013. LNCS, vol. 7918, pp. 107–126. Springer, Heidelberg (2013). doi:10.1007/978-3-642-38553-7_6

[FGO+13] Faugère, J.-C., Gauthier, V., Otmani, A., Perret, L., Tillich, J.-P.: A distinguisher for high rate McEliece cryptosystems. IEEE Trans. Inform. Theory 59(10), 6830–6844 (2013)

[FM08] Faure, C., Minder, L.: Cryptanalysis of the McEliece cryptosystem over hyperelliptic curves. In: Proceedings of the Eleventh International Workshop on Algebraic and Combinatorial Coding Theory, Pamporovo, Bulgaria, pp. 99–107, June 2008

[FOP+16] Faugère, J.-C., Otmani, A., Perret, L., de Portzamparc, F., Tillich, J.-P.: Folding alternant and Goppa Codes with non-trivial automorphism groups. IEEE Trans. Inform. Theory **62**(1), 184–198 (2016)

[Gab05] Gaborit, P.: Shorter keys for code based cryptography. In: Proceedings of the 2005 International Workshop on Coding and Cryptography (WCC 2005), Bergen, Norway, pp. 81–91, March 2005

[Gab08] Gabidulin, E.M.: Attacks and counter-attacks on the GPT public key cryptosystem. Des. Codes Cryptogr. **48**(2), 171–177 (2008)

[Gal63] Gallager, R.G.: Low Density Parity Check Codes. M.I.T. Press, Cambridge (1963)

[GM13] Gueye, C.T., Mboup, E.H.M.: Secure cryptographic scheme based on modified Reed Muller codes. Int. J. Secur. Appl. **7**(3), 55–64 (2013)

[GMRZ13] Gaborit, P., Murat, G., Ruatta, O., Zémor, G.: Low rank parity check codes and their application to cryptography. In: Proceedings of the Workshop on Coding and Cryptography (WCC 2013), Bergen, Norway (2013). www.selmer.uib.no/WCC2013/pdfs/Gaborit.pdf

[GO01] Gabidulin, E.M., Ourivski, A.V.: Modified GPT PKC with right scrambler. Electron. Notes Discrete Math. **6**, 168–177 (2001)

[Gop70] Goppa, V.D.: A new class of linear correcting codes. Problemy Peredachi Informatsii **6**(3), 24–30 (1970)

[GP13] Gabidulin, E., Pilipchuk, N.: GPT cryptosystem for information network security. In: International Conference on Information Society (i-Society 2013), no. 8, pp. 21–25 (2013)

[GP14] Gabidulin, E., Pilipchuk, N.: Modified GPT cryptosystem for information network security. Int. J. Inf. Secur. Res. **4**(8), 937–946 (2014)

[GPT91] Gabidulin, E.M., Paramonov, A.V., Tretjakov, O.V.: Ideals over a non-commutative ring and their application in cryptology. In: Davies, D.W. (ed.) EUROCRYPT 1991. LNCS, vol. 547, pp. 482–489. Springer, Heidelberg (1991). doi:10.1007/3-540-46416-6_41

[GRH09] Gabidulin, E., Rashwan, H., Honary, B.: On improving security of GPT cryptosystems. In: Proceedings of IEEE International Symposium on Information Theory (ISIT), pp. 1110–1114. IEEE (2009)

[GRS16] Gaborit, P., Ruatta, O., Schrek, J.: On the complexity of the rank syndrome decoding problem. IEEE Trans. Inf. Theory **62**(2), 1006–1019 (2016)

[GZ16] Gaborit, P., Zémor, G.: On the hardness of the decoding and the minimum distance problems for rank codes. IEEE Trans. Inf. Theory **62**(12), 7245–7252 (2016)

[HMR15] Horlemann-Trautmann, A.-L., Marshall, K., Rosenthal, J.: Extension of overbeck's attack for gabidulin based cryptosystems. CoRR, abs/1511.01549 (2015)

[HPS98] Hoffstein, J., Pipher, J., Silverman, J.H.: NTRU: a ring-based public key cryptosystem. In: Buhler, J.P. (ed.) ANTS 1998. LNCS, vol. 1423, pp. 267–288. Springer, Heidelberg (1998). doi:10.1007/BFb0054868

[HSEA14] Hooshmand, R., Koochak Shooshtari, M., Eghlidos, T., Aref, M.R.: Reducing the key length of McEliece cryptosystem using polar codes. In: 2014 11th International ISC Conference on Information Security and Cryptology (ISCISC), pp. 104–108. IEEE (2014)

[HTMR16] Horlemann-Trautmann, A.-L., Marshall, K., Rosenthal, J.: Considerations for rank-based cryptosystems. In: 2016 IEEE International Symposium on Information Theory (ISIT), pp. 2544–2548. IEEE (2016)

[HvMG13] Heyse, S., von Maurich, I., Güneysu, T.: Smaller keys for code-based cryptography: QC-MDPC McEliece implementations on embedded devices. In: Bertoni, G., Coron, J.-S. (eds.) CHES 2013. LNCS, vol. 8086, pp. 273–292. Springer, Heidelberg (2013). doi:10.1007/978-3-642-40349-1_16

[ISR60] Solomon, G., Reed, I.S.: Polynomial codes over certain finite fields. J. Soc. Industr. Appl. Math. **8**(2), 300–304 (1960)

[JM96] Janwa, H., Moreno, O.: McEliece public key cryptosystems using algebraic-geometric codes. Des. Codes Cryptogr. **8**(3), 293–307 (1996)

[KKM+17] Kudekar, S., Kumar, S., Mondelli, M., Pfister, H.D., Sasoglu, E., Urbanke, R.: Reed-muller codes achieve capacity on erasure channels. IEEE Trans. Inf. Theory **PP**(99), 1 (2017)

[KMP14] Kiltz, E., Masny, D., Pietrzak, K.: Simple chosen-ciphertext security from low-noise LPN. In: Krawczyk, H. (ed.) PKC 2014. LNCS, vol. 8383, pp. 1–18. Springer, Heidelberg (2014). doi:10.1007/978-3-642-54631-0_1

[LDW94] Li, Y.X., Deng, R.H., Wang, X.M.: On the equivalence of McEliece's and Niederreiter's public-key cryptosystems. IEEE Trans. Inform. Theory **40**(1), 271–273 (1994)

[LJ12] Löndahl, C., Johansson, T.: A new version of McEliece PKC based on convolutional codes. In: Chim, T.W., Yuen, T.H. (eds.) ICICS 2012. LNCS, vol. 7618, pp. 461–470. Springer, Heidelberg (2012). doi:10.1007/978-3-642-34129-8_45

[Loi10] Loidreau, P.: Designing a rank metric based McEliece cryptosystem. In: Sendrier, N. (ed.) PQCrypto 2010. LNCS, vol. 6061, pp. 142–152. Springer, Heidelberg (2010). doi:10.1007/978-3-642-12929-2_11

[LS01] Loidreau, P., Sendrier, N.: Weak keys in the McEliece public-key cryptosystem. IEEE Trans. Inform. Theory **47**(3), 1207–1211 (2001)

[LT13] Landais, G., Tillich, J.-P.: An efficient attack of a McEliece cryptosystem variant based on convolutional codes. In: Gaborit, P. (ed.) PQCrypto 2013. LNCS, vol. 7932, pp. 102–117. Springer, Heidelberg (2013). doi:10.1007/978-3-642-38616-9_7

[Mac99] MacKay, D.J.C.: Good error-correcting codes based on very sparse matrices. IEEE Trans. Inf. Theory **45**(2), 399–431 (1999)

[MB09] Misoczki, R., Barreto, P.S.L.M.: Compact McEliece keys from goppa codes. In: Jacobson, M.J., Rijmen, V., Safavi-Naini, R. (eds.) SAC 2009. LNCS, vol. 5867, pp. 376–392. Springer, Heidelberg (2009). doi:10.1007/978-3-642-05445-7_24

[McE78] McEliece, R.J.: A public-key system based on algebraic coding theory, pp. 114–116. Jet Propulsion Lab, DSN Progress Report 44 (1978)

[MCT16] Márquez-Corbella, I., Tillich, J.-P.: Using Reed-Solomon codes in the $(u|u + v)$ construction and an application to cryptography. In: Proceedings of IEEE International Symposium on Information Theory (ISIT), pp. 930–934 (2016). arXiv:1601:08227

[Mit51] Mitani, N.: On the transmission of numbers in a sequential computer. National Convention of the Institute of Electrical Communication Engineers of Japan, November 1951

[MOG15] Maurich, I.V., Oder, T., Güneysu, T.: Implementing QC-MDPC McEliece encryption. ACM Trans. Embed. Comput. Syst. **14**(3), 44:1–44:27 (2015)

[MRAS00] Monico, C., Rosenthal, J., Shokrollahi, A.A.: Using low density parity check codes in the McEliece cryptosystem. In: Proceedings of IEEE International Symposium on Information Theory (ISIT), Sorrento, Italy, p. 215 (2000)

[MS86] MacWilliams, F.J., Sloane, N.J.A.: The Theory of Error-Correcting Codes, 5th edn. North-Holland, Amsterdam (1986)

[MS07] Minder, L., Shokrollahi, A.: Cryptanalysis of the sidelnikov cryptosystem. In: Naor, M. (ed.) EUROCRYPT 2007. LNCS, vol. 4515, pp. 347–360. Springer, Heidelberg (2007). doi:10.1007/978-3-540-72540-4_20

[MTSB13] Misoczki, R., Tillich, J.-P., Sendrier, N., Barreto, P.S.L.M.: MDPC-McEliece: new McEliece variants from moderate density parity-check codes. In: Proceedings of IEEE International Symposium on Information Theory (ISIT), pp. 2069–2073 (2013)

[Mul54] Muller, D.E.: Application of boolean algebra to switching circuit design, to error detection. Trans. I.R.E. Prof. Group Electron. Comput. **EC-3**(3), 6–12 (1954)

[Nie86] Niederreiter, H.: Knapsack-type cryptosystems and algebraic coding theory. Probl. Control Inf. Theory **15**(2), 159–166 (1986)

[OB09] Ouzan, S., Be'ery, Y.: Moderate-density parity-check codes. arXiv preprint (2009). arXiv:0911.3262

[OTD08] Otmani, A., Tillich, J.-P., Dallot, L.: Cryptanalysis of McEliece cryptosystem based on quasi-cyclic LDPC codes. In: Proceedings of First International Conference on Symbolic Computation and Cryptography, Beijing, China, 28–30 April 2008, pp. 69–81. LMIB Beihang University (2008)

[OTK15] Otmani, A., Kalachi, H.T.: Square code attack on a modified sidelnikov cryptosystem. In: El Hajji, S., Nitaj, A., Carlet, C., Souidi, E.M. (eds.) C2SI 2015. LNCS, vol. 9084, pp. 173–183. Springer, Cham (2015). doi:10.1007/978-3-319-18681-8_14

[OTKN16] Otmani, A., Talé-Kalachi, H., Ndjeya, S.: Improved cryptanalysis of rank metric schemes based on Gabidulin codes. CoRR, abs/1602.08549 (2016)

[Ove08] Overbeck, R.: Structural attacks for public key cryptosystems based on Gabidulin codes. J. Cryptol. **21**(2), 280–301 (2008)

[Per12] Persichetti, E.: Compact McEliece keys based on quasi-dyadic Srivastava codes. J. Math. Cryptol. **6**(2), 149–169 (2012)

[PR97] Petrank, E., Roth, R.: Is code equivalence easy to decide? IEEE Trans. Inform. Theory **43**(5), 1602–1604 (1997)

[Ree54] Reed, I.S.: A class of multiple-error-correcting codes and the decoding scheme. IRE Trans. IT **4**, 38–49 (1954)

[RGH10] Rashwan, H., Gabidulin, E., Honary, B.: A smart approach for GPT cryptosystem based on rank codes. In: Proceedings of IEEE International Symposium on Information Theory (ISIT), pp. 2463–2467. IEEE (2010)

[RGH11] Rashwan, H., Gabidulin, E., Honary, B.: Security of the GPT cryptosystem and its applications to cryptography. Secur. Commun. Netw. **4**(8), 937–946 (2011)

[RSA78] Rivest, R.L., Shamir, A., Adleman, L.M.: A method for obtaining digital signatures and public-key cryptosystems. Commun. ACM **21**(2), 120–126 (1978)

[Sen94] Sendrier, N.: On the structure of a randomly permuted concatenated code. In: EUROCODE 1994, pp. 169–173 (1994)

[Sen98] Sendrier, N.: On the concatenated structure of a linear code. Appl. Algebra Eng. Commun. Comput. (AAECC) **9**(3), 221–242 (1998)

[Sen00] Sendrier, N.: Finding the permutation between equivalent linear codes: the support splitting algorithm. IEEE Trans. Inf. Theory **46**(4), 1193–1203 (2000)

[Sha48] Shannon, C.E.: A mathematical theory of communication. Bell Syst. Tech. J. **27**(3), 379–423 (1948)

[Sho94] Shor, P.W.: Algorithms for quantum computation: discrete logarithms and factoring. In: Goldwasser, S. (ed.) FOCS, pp. 124–134 (1994)

[Sid94] Sidelnikov, V.M.: A public-key cryptosytem based on Reed-Muller codes. Discr. Math. Appl. **4**(3), 191–207 (1994)

[SK14] Shrestha, S.R., Kim, Y.-S.: New McEliece cryptosystem based on polar codes as a candidate for post-quantum cryptography. In: 2014 14th International Symposium on Communications and Information Technologies (ISCIT), pp. 368–372. IEEE (2014)

[SS92] Sidelnikov, V.M., Shestakov, S.O.: On the insecurity of cryptosystems based on generalized Reed-Solomon codes. Discr. Math. Appl. **1**(4), 439–444 (1992)

[SS96] Sipser, M., Spielman, D.A.: Expander codes. IEEE Trans. Inf. Theory **42**, 1710–1722 (1996)

[SSMS09] Shoufan, A., Strenzke, F., Molter, H.G., Stöttinger, M.: A timing attack against patterson algorithm in the McEliece PKC. In: Lee, D., Hong, S. (eds.) ICISC 2009. LNCS, vol. 5984, pp. 161–175. Springer, Heidelberg (2010). doi:10.1007/978-3-642-14423-3_12

[STM+08] Strenzke, F., Tews, E., Molter, H.G., Overbeck, R., Shoufan, A.: Side channels in the McEliece PKC. In: Buchmann, J., Ding, J. (eds.) PQCrypto 2008. LNCS, vol. 5299, pp. 216–229. Springer, Heidelberg (2008). doi:10.1007/978-3-540-88403-3_15

[Str10a] Strenzke, F.: A timing attack against the secret permutation in the McEliece PKC. In: Sendrier, N. (ed.) PQCrypto 2010. LNCS, vol. 6061, pp. 95–107. Springer, Heidelberg (2010). doi:10.1007/978-3-642-12929-2_8

[Str10b] Strenzke, F.: A smart card implementation of the McEliece PKC. In: Samarati, P., Tunstall, M., Posegga, J., Markantonakis, K., Sauveron, D. (eds.) WISTP 2010. LNCS, vol. 6033, pp. 47–59. Springer, Heidelberg (2010). doi:10.1007/978-3-642-12368-9_4

[Str13] Strenzke, F.: Timing attacks against the syndrome inversion in code-based cryptosystems. In: Gaborit, P. (ed.) PQCrypto 2013. LNCS, vol. 7932, pp. 217–230. Springer, Heidelberg (2013). doi:10.1007/978-3-642-38616-9_15

[UG14] Urivskiy, A., Gabidulin, E.: On the equivalence of different variants of the GPT cryptosystem, no. 3, pp. 95–97. IEEE (2014)

[Wan16] Wang, Y.: Quantum resistant random linear code based public key encryption scheme rlce. In: 2016 IEEE International Symposium on Information Theory (ISIT), pp. 2519–2523. IEEE (2016)

[Wie06a] Wieschebrink, C.: An attack on a modified niederreiter encryption scheme. In: Yung, M., Dodis, Y., Kiayias, A., Malkin, T. (eds.) PKC 2006. LNCS, vol. 3958, pp. 14–26. Springer, Heidelberg (2006). doi:10.1007/11745853_2

[Wie06b] Wieschebrink, C.: Two NP-complete problems in coding theory with an application in code based cryptography. In: Proceedings of IEEE International Symposium on Information Theory (ISIT), pp. 1733–1737 (2006)

[Wie09] Wieschebrink, C.: Cryptanalysis of the Niederreiter public key scheme based on GRS subcodes. IACR Cryptology ePrint Archive, Report 2009/452 (2009)

New Algorithm for Modeling S-box in MILP Based Differential and Division Trail Search

Yu Sasaki$^{(\boxtimes)}$ and Yosuke Todo

NTT Secure Platform Laboratories,
3-9-11 Midori-cho, Musashino-shi, Tokyo 180-8585, Japan
{sasaki.yu,todo.yosuke}@lab.ntt.co.jp

Abstract. This paper studies an automated differential-trail search against block ciphers in which the problem of finding the optimal trail is converted to one of finding the optimal solution in a mixed-integer-linear programming (MILP). The most difficult part is representing differential properties of an S-box, known as differential distribution table (DDT), with a system of inequalities. Previous work builds the system by using a general-purpose mathematical tool, SAGE Math. However, the generated system for general-purpose contains a lot of redundant inequalities for the purpose of differential-trail search, thus inefficient. Hence, an auxiliary algorithm was introduced to minimize the number of inequalities by hoping that it minimizes the runtime to solve the MILP. This paper proposes a new algorithm to improve this auxiliary algorithm. The main advantage is that while the previous algorithm does not ensure the minimum number of inequalities, the proposed algorithm does ensure it. Moreover it enables the users to choose the number of inequalities in the system. In addition, this paper experimentally shows that the above folklore "minimizing the number of inequalities minimizes the runtime" is not always correct. The proposed algorithm can also be used in the MILP-based division-trail search, which evaluates the bit-based division property for integral attacks.

Keywords: Differential trail · Division trail · Automated search tool · S-box · Mixed integer linear programming · Greedy algorithm

1 Introduction

Symmetric-key primitives like block ciphers are one of the most fundamental parts of cryptography. A lot of new designs have been proposed continuously to investigate good designs achieving both of high security and efficiency.

Differential cryptanalysis developed by Biham and Shamir [1] is one of the most generic cryptanalytic approaches. It is now almost mandatory for designers to evaluate security against differential cryptanalysis. Resistance against differential cryptanalysis can increase by using non-linear operations during the computation. The S-box, predefined substitution table, is a typical design choice to introduce non-linearity. A popular approach for evaluating differential cryptanalysis against an S-box based design is as follows.

© Springer International Publishing AG 2017
P. Farshim and E. Simion (Eds.): SecITC 2017, LNCS 10543, pp. 150–165, 2017.
https://doi.org/10.1007/978-3-319-69284-5_11

1. Evaluate the maximum probability of differential propagation in a single S-box denoted by p_S^{max}. Namely for S-box $S : \{0,1\}^n \mapsto \{0,1\}^n$, p_S^{max} is defined as

$$p_S^{max} \triangleq \max_{\Delta_i, \Delta_o} \left\{ \frac{\#x \in \{0,1\}^n | S(x) \oplus S(x \oplus \Delta_i) = \Delta_o}{2^n} \right\}. \tag{1}$$

2. Search for the differential propagation pattern for the entire algorithm that minimizes the number of S-boxes with difference, denoted by active S-boxes. This process is known as differential trail search. Let N_{AS} be the lower bound of the number of active S-boxes.
3. The probability of a differential trail is upper-bounded by $N_{AS} \times P_S^{max}$.

The differential trail search is the most difficult part. Some designs, e.g. AES, adopt a clever computation structure such that the minimum number of active S-boxes can be calculated easily, which is called wide trails strategy. However, it cannot be applied to any algorithm especially for complicated computation structure.

In 2011, Mouha et al. proposed an automated differential path search method by using mixed-integer-linear programming (MILP) [2], which generates lower bounds of the number of active S-boxes. At that time, the search tool could not consider the differential property of the S-box, thus could not apply to bit-oriented ciphers such as PRESENT [3] and LS-designs [4].

This restriction was later solved by Siwei et al. [5,6], which describes the possible and impossible differential propagations of the S-box with a system of inequalities. The goal of our research is improving the S-box description of [5,6], thus we explain their method deeply.

S-box Modeling in Previous Work. Suppose that the S-box we want to analyze is the 4-bit S-box defined in Table 1. Its differential distribution table (DDT), i.e. the value of $\#x$ in Eq. (1) for each of (Δ_i, Δ_o), is given in Table 2. Also suppose that x_3, x_2, x_1, x_0 are four binary variables in which $x_j = 0$ and $x_j = 1$ for $j \in \{0, 1, 2, 3\}$ represent that j-th input bit to the S-box is inactive and active, respectively. Similarly y_3, y_2, y_1, y_0 are four binary variables to represent the active status of the S-box output. Here the goal is building a system of linear inequalities with respect to $(x_3, x_2, x_1, x_0, y_3, y_2, y_1, y_0)$ so that the solution space matches the non-zero entries in DDT. From Table 2, input difference 0x1 never propagates to output difference 0x2. Thus $(x_3, x_2, x_1, x_0, y_3, y_2, y_1, y_0) =$ (00010010) must be removed from the search range. On the other hand $\Delta_i = $ 0x1 can propagate to $\Delta_o = $ 0x1 with non-zero probability. Thus $(x_3, x_2, x_1, x_0, y_3, y_2, y_1, y_0) = $ (00010001) must be included in the solution space. Similarly, from $2^8 = 256$ patterns of $(x_3, x_2, x_1, x_0, y_3, y_2, y_1, y_0)$, we now have a set of patterns that must be excluded from the solution space (150 entries with 0 in Table 2).

The above can be generalized to the following problem; For a given subspace $\mathcal{R} \subset \mathbb{F}_2^n$, we want to build a system of inequalities so that $\mathbb{F}_2^n - \mathcal{R}$ matches a solution space. Two approaches are known to build such a system of inequalities.

SAGE Math: Due to this highly generic problem, a mathematical tool, *SAGE Math*, equips the function to generate such a system of inequalities. Namely it takes as input a subset of $\mathbb{F}_2^n$, and returns a system of inequalities with the form of $\alpha_0^i x_0 + \alpha_1^i x_1 + \cdots + \alpha_{n-1}^i x_{n-1} + \alpha_n^i \geq 0$, where α_j^i is an integer coefficient for the ith inequality. Here, SAGE Math assumes an *integer programming*, namely it regards that each of $x_0, x_1, \cdots, x_{n-1}$ is an integer variable. However, the differential-trail search is *01-integer programming*, namely each variable takes only 0 or 1. As a result, the system generated by SAGE Math is too heavy and contains a lot of overlap when the value of each variable is limited to 0 or 1. Hence, it requires an auxiliary code to exclude the overlapping inequalities.

Logical Computation Model: It is also possible to directly build a system of inequalities for 01-integer programming without using the existing mathematical tool. Suppose that we want to remove $(x_3, x_2, x_1, x_0, y_3, y_2, y_1, y_0) = (00010010)$ from the solution space. Then, we can specify $x_3 + x_2 + x_1 - x_0 + y_3 + y_2 - y_1 + y_0 \geq -1$, which removes this pattern while keeps all the other patterns in the solution space. Moreover, suppose that (00010011) should also be removed. In this case, two patterns (00010010) and (00010011) can be removed together, *i.e.* $(0001001*)$ can be removed by $x_3 + x_2 + x_1 - x_0 + y_3 + y_2 - y_1 \geq -1$. Such a merged form yields the overlap issue in the logical computation model as well as SAGE Math. Namely, if $(0001001*)$ is removed, we no longer need to remove (00010010) and (00010011) one by one.

Table 1. An example of S-box

x	0	1	2	3	4	5	6	7	8	9	A	B	C	D	E	F
$S(x)$	4	8	7	1	9	3	2	E	0	B	6	F	A	5	D	C

In both approaches, a straightforward system of inequalities to remove $\mathcal{R}$ contains too many overlaps, e.g. SAGE Math generates more than 300 inequalities for 4-bit S-box while only 20 to 30 inequalities are sufficient to strictly define $\mathbb{F}_2^4 - \mathcal{R}$. Previous work [5,6] argued that *minimizing the number of inequalities* is important to minimize the runtime to solve MILP and proposed using the *greedy algorithm* to exclude redundant inequalities. More precisely, the greedy algorithm picks an inequality which excludes more elements in $\mathcal{R}$ from the solution space than any other inequalities. Then the same procedure continues for the remaining inequalities and elements in $\mathcal{R}$ until all the elements in $\mathcal{R}$ are excluded from the solution space. For the sake of simplicity, we call the algorithm to minimize the number of inequalities *reduction algorithm*.

The use of the greedy algorithm has two drawbacks. First, it does not guarantee the optimal choice. Second, there are many inequalities having a tied score during the execution of the greedy algorithm but the authors [5,6] did not specify the choice of this case, thus their algorithm does not have reproducibility.

Table 2. Differential Distribution Table (DDT)

Δ_i

	0	1	2	3	4	5	6	7	8	9	a	b	c	d	e	f
0	16	0	0	0	0	0	0	0	0	0	0	0	0	0	0	0
1	0	2	0	0	0	0	2	0	0	2	2	2	4	0	0	2
2	0	0	0	2	2	0	2	2	0	4	0	2	0	2	0	0
3	0	2	0	0	0	2	2	2	2	0	0	0	0	2	0	4
4	0	0	0	2	0	2	0	0	0	0	2	4	0	2	2	2
5	0	4	2	2	0	2	0	2	0	2	2	0	0	0	0	0
6	0	0	2	0	0	0	4	2	0	0	2	0	2	2	2	0
7	0	0	0	2	2	2	2	0	2	0	4	0	2	0	0	0
8	0	2	2	4	2	0	2	0	0	0	0	0	0	0	2	2
9	0	0	0	2	0	0	0	2	4	2	0	0	2	0	2	2
a	0	0	2	0	2	0	0	4	2	0	2	2	0	0	0	2
b	0	2	0	0	2	2	0	2	0	0	0	2	2	0	4	0
c	0	2	0	0	2	0	0	0	2	2	2	0	0	4	2	0
d	0	2	4	2	0	0	0	0	2	0	0	2	2	2	0	0
e	0	0	2	0	4	2	0	0	0	2	0	0	2	2	0	2
f	0	0	2	0	0	4	2	0	2	2	0	2	0	0	2	0

Δ_o labels the rows.

Indeed, we could not reproduce the same result as [5,6] and this is a part of our motivation to develop a new algorithm.

Our Contributions. The current paper proposes a new reduction algorithm when the S-box is modeled with MILP. More precisely, we first use either SAGE Math or logical computation model to obtain a large system of inequalities including redundant ones. Then as the reduction algorithm, we use our new algorithm instead of the greedy algorithm in previous work. Interestingly, the new reduction algorithm is based on MILP. Namely, we convert the problem of minimizing the number of inequalities for excluding $\mathcal{R}$ to the problem of minimizing the sum of involved inequalities in some MILP problem.

The new algorithm inherits the advantage of the MILP such that the optimal solution is obtained. Previous work [6] listed the number of inequalities which they obtained by applying the greedy algorithm to 4-bit S-boxes in various ciphers; Kline [7], Piccolo [8], TWINE [9], PRINCE [10], MIBS [11], PRESENT [3], LED [12], LBlock [13], Serpent [14]. For comparison, we also apply the new reduction algorithm to those S-boxes. The results are shown in Table 3. The new reduction algorithm finds a smaller set of inequalities for all S-boxes but for TWINE. For completeness, we also apply our reduction algorithm to several other ciphers recently designed, which includes LILLIPUT [15], Midori [16], Minalpher [17], RECTANGLE [18], SKINNY [19].

With the new reduction algorithm, the user can choose the number of equalities which will be incorporated into the system. This property ensures the reproducibility of the results. Namely, every user can obtain the system with the same number of equalities by simply running the existing MILP solver.

This feature enables us to perform experiments of solving the same problem with various numbers of inequalities to represent the S-box, which would reveal the relationship between the number of inequalities and the runtime for the entire differential-trail search problem. The results show that when the number

Table 3. Number of inequalities to exclude $\mathcal{R}$ for various 4-bit S-boxes

Sbox	#inequalities			Sbox	#inequalities		
	SAGE Math	Previous	Ours		SAGE Math	Previous	Ours
Kline	311	22	21	LBlock S6	205	27	24
Piccolo	202	23	21	LBlock S7	205	27	24
TWINE	324	23	23	LBlock S8	205	28	24
PRINCE	300	26	22	LBlock S9	205	27	24
MIBS	378	27	23	Serpent S0	327	23	21
PRESENT/LED	327	22	21	Serpent S1	327	24	21
LBlock S0	205	28	24	Serpent S2	325	25	21
LBlock S1	205	27	24	Serpent S3	368	31	27
LBlock S2	205	27	24	Serpent S4	321	26	23
LBlock S3	205	27	24	Serpent S5	321	25	23
LBlock S4	205	28	24	Serpent S6	327	22	21
LBlock S5	205	27	24	Serpent S7	368	30	27
Lilliput	324	—	23	Minalpher	338	—	22
Midori S0	239	—	21	RECTANGLE	267	—	21
Midori S1	367	—	22	SKINNY	202	—	21

of inequalities for each S-box is too small, the runtime for the entire algorithm is significantly longer, which runs contrary to the previous belief that minimizing the number of inequalities for each S-box is the best.

Paper Outline. The remaining of this paper is structured as follows. Section 2 explains how to search for differential trails with MILP. Section 3 explains the previous reduction algorithm based on the greedy algorithm. Section 4 presents our new reduction algorithm based on MILP. Section 5 shows the experiments that minimizing the number of inequalities is not the best strategy. Section 6 discusses the application to the division-trail search.

2 Differential Trail Search with MILP

In this section, we explain how the problem of finding the best differential trail is converted to the problem of solving MILP.

Suppose that we build an MILP model for a block cipher whose block size is b bits and the number of rounds is r. Let $s_0, s_1, \ldots, s_{b-1}$ be b bits of the plaintext. Similarly, let $s_{b*j+0}, s_{b*j+1}, \ldots, s_{b*j+b-1}$, $j = 1, 2, \cdots, r$ be b bits of the state after round j, thus $s_{b*r+0}, s_{b*r+1}, \ldots, s_{b*r+b-1}$ are b bits of the ciphertext.

To make an MILP model, binary variables, $x_i \in \{0, 1\}$ where $i = 0, 1, \cdots, b * r + b - 1$, are firstly introduced to represent whether the bit s_i is active or not. $x_i = 0$ indicates that s_i is inactive while $x_i = 1$ indicates that s_i is active.

Secondly, the valid differential propagation is modeled by using inequalities among variables x_i. It is more convenient to discuss an example. Here, we consider the toy example in Fig. 1, in which the block size is 8 bits and round function consists of application of 4-bit S-box for the top 4 bits and the bottom 4 bits, xoring the top 4 bits to the bottom, and swap those 4 bits.

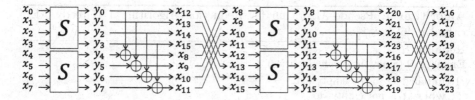

Fig. 1. Binary variables for 2-round toy cipher

Each input bit to the first round and the second round is modeled by x_0-$\hat{x}_7$ and x_8-x_{15}, respectively. The last swap is the permutation of bit positions, thus the state before the swap in the first round can be described with x_8-x_{15}. Due to the requirement of the complex method, new binary variables y_0-y_7 are introduced to represent the active status of each output bit from S-box. Figure 1 includes those binary variables.

Model for Linear Part. Swap operation, or any other permutation of bit positions, does not require any inequalities. When we refer to top 4 bits before the swap operation in the first round, we can directly refer to $x_{12}, x_{13}, x_{14}, x_{15}$.

MILP accepts using equation to construct the model. Thus the relation $y_0 = x_{12}$, $y_1 = x_{13}$, $y_2 = x_{14}$, and $y_3 = x_{15}$ can be modeled directly.

Then we consider modeling valid differential propagation for 2-bit xor, i.e. $a \oplus b = c$. We can exclude impossible propagation patterns one by one with one inequality. The impossible patterns are $(a, b, c) = (1, 0, 0), (0, 1, 0), (0, 0, 1), (1, 1, 1)$.

- $(a, b, c) = (1, 0, 0)$ can be removed by $-a + b + c \geq 0$. Indeed, $(a, b, c) = (1, 0, 0)$ does not satisfy this inequality, and all the other patterns remain in the solution space.
- $(a, b, c) = (1, 0, 0)$ can be removed by $a - b + c \geq 0$.
- $(a, b, c) = (0, 1, 0)$ can be removed by $a + b - c \geq 0$.
- $(a, b, c) = (1, 1, 1)$ can be removed by $-a - b - c \geq -2$.

In the end, by using the above four inequalities, the valid differential propagation for $a \oplus b = c$ is modeled. By replacing (a, b, c) with (y_0, y_4, x_8), (y_1, y_5, x_9), (y_2, y_6, x_{10}), and (y_3, y_7, x_{11}), the xor operations of the first round can be modeled with $4 * 4 = 16$ inequalities.

Model for 4-Bit S-box. The model for S-box has already been explained in Sect. 1. In this paper, we focus on using SAGE Math. The user first analyzes DDT of the target S-box and makes a list of valid patterns of $(x_3 x_2 x_1 x_0 y_3 y_2 y_1 y_0)$. The user then passes the list to SAGE Math to obtain a system of inequalities in which the solution space corresponds to the given patterns. The system looks as follows, which is a result of simulating DDT in Table 2.

Line 1: An inequality $(-1, 0, 0, 0, 0, 0, 0, 0)x + 1 >= 0$
Line 2: An inequality $(0, -1, 0, 0, 0, 0, 0, 0)x + 1 >= 0$
Line 3: An inequality $(0, 0, -1, 0, 0, 0, 0, 0)x + 1 >= 0$

Algorithm 1. Pseudo-Algorithm for Reduction Algorithm 1 (Greedy Algorithm) in [6]

Require: $\mathcal{H}$, $\mathcal{X}$
Ensure: $\mathcal{O}$ (a new list of inequalities)

1: Initialize $\mathcal{O}$ to the empty list.
2: **while** $\mathcal{X}$ is not empty **do**
3:     Pick up an inequality in $\mathcal{H}$ which maximizes the number of removed impossible patterns in $\mathcal{X}$.
4:     Add the inequality to $\mathcal{O}$.
5:     Erase the inequality from $\mathcal{H}$.
6:     Erase the removed impossible patterns from $\mathcal{X}$.
7: **end while**
8: **return** $\mathcal{O}$

Line 4: An inequality $(0, 0, 0, -1, 0, 0, 0, 0)x + 1 >= 0$
Line 5: An inequality $(0, 0, 0, 0, -1, 0, 0, 0)x + 1 >= 0$
Line 6: An inequality $(0, -1, -1, -1, -1, -1, 0, -1)x + 5 >= 0$
Line 7: An inequality $(-1, 0, -1, -1, -1, -1, 0, -1)x + 5 >= 0$

$$\vdots$$

Line 323: An inequality $(-1, -1, -1, 0, -1, 1, -1, 1)x + 4 >= 0$
Line 324: An inequality $(-1, -2, -1, -3, -1, -3, -2, -3)x + 13 >= 0$

Each line indicates 8 coefficients for each variable and the constant value. For example, the last line denotes $-x_3 - 2x_2 - x_1 - 3x_0 - y_3 - 3y_2 - 2y_1 - 3y_0 + 13 \geq 0$. Thus each S-box can be modeled with 324 inequalities, and all the S-layers of the r-round toy cipher can be modeled with $324 \times 2 \times r$ inequalities.

A natural concern is that using 324 inequalities per S-box is too costly. Indeed, the system generated by SAGE Math assumes that each variable can take any integer, while they take only 0 or 1 in differential search. Thus, a lot of inequalities are actually unnecessary. To reduce the number of inequalities, Siwei et al. [5,6] introduced the greedy algorithm as follows.

Let $\mathcal{H}$ be a list of inequalities generated by SAGE Math. Let $\mathcal{X}$ be a list of impossible differential propagation patterns to be removed, which initialized to $\mathcal{R}$. Their algorithm generates a new list of inequalities, $\mathcal{O}$, which is much smaller than $\mathcal{H}$. Intuitively, $\mathcal{O}$ is first initialized to the empty set, and they add inequality of $\mathcal{H}$ to $\mathcal{O}$ one by one so that at each timing the number of removed impossible patterns of $\mathcal{X}$ is maximized. The pseudo-algorithm is given in Algorithm 1.

In this paper, we call the algorithm to reduce the number of inequalities *reduction algorithm*, which is the main object of this paper.

Solving the System. After the system of inequalities is generated, we need to find an optimal solution. This part is generally done by using existing software to solve MILP. A number of MILP solvers are available such as Gurobi Optimizer [20], SCIP [21] and CPLEX Optimization Studio [22]. Some of them are commercial products but many of them offer a free license for academic organizations.

3   Problems of Reduction Algorithm in Previous Work

In general, the greedy algorithm does not guarantee the optimality of the solution, and this actually applies to the problem of finding minimal representation to describe DDT. We provide a counterexample to demonstrate this fact. Let $\mathcal{R}$ be a set of 8 elements in $\mathbb{F}_2^4$ such that $\mathcal{R} \triangleq \{0001, 0101, 0111, 0110, 1100, 1101, 1111, 1011\}$, and here the goal is finding a minimal representation of $\mathcal{R}$. Figure 2(a) represents $\mathcal{R}$ in the Karnaugh map.

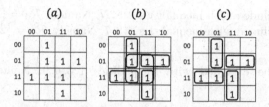

Fig. 2. An example that greedy algorithm does not minimize constraints

In Fig. 2(b), the greedy algorithm is applied. Firstly, four elements in the center $(0101, 0111, 1101, 1111)$ are simplified to $*1*1$, and then four elements remain. As a result, $\mathcal{R}$ is represented by five elements $\{*1*1, 0*01, 110*, 011*, 1*11\}$. However, it is easy to see that the last four elements are sufficient to cover $\mathcal{R}$ (Fig. 2(c)), which shows that the greedy algorithm is not optimal.

4   New Reduction Algorithm

In this section, we describe our new reduction algorithm to search for the representation of an n-bit S-box with the minimum number of inequalities. We assume that the following analysis finished before we run the reduction algorithm.

- DDT (with 2^{2n} entries) is computed. Let $x_{n-1}, x_{n-2}, \cdots, x_0$ and $y_{n-1}, y_{n-2}, \cdots, y_0$ be n binary variables to denote whether each of input and output bit is active or not, respectively. Then the attacker obtains a set of all impossible differential propagation patterns of $(x_{n-1}, x_{n-2}, \cdots, x_0, y_{n-1}, y_{n-2}, \cdots, y_0)$. This set is denoted by $\mathcal{R}$. Suppose that there are $|\mathcal{R}|$ elements in $\mathcal{R}$. We denote each element of $\mathcal{R}$ by $R_0, R_1, R_2, \cdots, R_{|\mathcal{R}|-1}$.
- A large size of the system of inequalities to represent $\mathbb{F}_2^{2n} - \mathcal{R}$ is obtained by using either SAGE Math or the logical computation model. We denote the number of inequalities before the reduction algorithm is applied by N.

In the end, we know $|\mathcal{R}|$ patterns that must be excluded from the solution space and we know N inequalities whose intersection achieves $\mathbb{F}_2^{2n} - \mathcal{R}$ but containing many redundant inequalities.

Overview. The overview of our reduction algorithm is as follows. First, for each impossible pattern R_i, we compute which inequalities can exclude R_i from the solution space. Second, for each R_i we make a constraint such that R_i must be excluded from the solution space by at least 1 inequality. Finally, under these constraints we minimize the number of inequalities.

Initial Process. We first perform a small pre-process. For each R_i, we check which of N inequalities exclude the pattern from the solution space. Let $\overline{\mathcal{R}}_i$ be a set of inequalities that can exclude R_i. For example, we consider the situation summarized in Table 4, which indicates as follows.

- R_0 can be excluded with inequalities $2, 8, N$. Namely, $\overline{\mathcal{R}}_0 = \{2, 8, N\}$.
- R_1 can be excluded with inequalities $2, 3, 7$. Namely, $\overline{\mathcal{R}}_1 = \{2, 3, 7\}$.

$$\vdots$$

- $R_{|\mathcal{R}|-1}$ can be excluded with inequalities $1, 3, 4, 9$. Namely, $\overline{\mathcal{R}}_{|\mathcal{R}|-1} = \{1, 3, 4, 9\}$.

Table 4. Example of precomputation analysis

| | Patterns in $\mathcal{R}$ | | | | | | | | | | | |
	R_0	R_1	R_2	R_3	R_4	R_5	R_6	R_7	R_8	R_9	$\cdots$	$R_{|\mathcal{R}|-1}$
Inequality 1	0	0	0	1	1	0	1	0	0	0	$\cdots$	1
Inequality 2	1	1	0	1	0	0	0	0	0	1	$\cdots$	0
Inequality 3	0	1	1	0	0	1	0	0	1	0	$\cdots$	1
Inequality 4	0	0	0	0	0	0	1	0	0	0	$\cdots$	1
Inequality 5	0	0	0	1	0	0	0	0	1	0	$\cdots$	0
Inequality 6	0	0	1	0	0	0	0	0	0	0	$\cdots$	0
Inequality 7	0	1	0	0	1	1	0	0	0	0	$\cdots$	0
Inequality 8	1	0	0	1	0	0	0	1	0	0	$\cdots$	0
Inequality 9	0	0	1	0	0	0	0	1	0	0	$\cdots$	1
$\vdots$					$\vdots$							
Inequality N	1	0	0	1	0	0	1	0	1	1	$\cdots$	0

Declaration of Variables. In this MILP, we use N binary variables $z_1, z_2, \cdots, z_N$, in which $z_i = 1$ denotes that inequality i is included in the system and $z_i = 0$ denotes that inequality i will not be used in the system.

Objective Function. The goal is minimizing the number of inequalities adopted in the system, which can be represented by

$$\text{minimize} \sum_{i=1}^{N} z_i. \tag{2}$$

Constraints. The only constraint we need is ensuring that each impossible pattern is removed with at least one inequality. Thus we have $|\mathcal{R}|$ constraints in which the sum of z_i with $i \in \overline{\mathcal{R}_i}$ is greater than or equal to 1. With the above example, we have the following constraints.

$$z_2 + z_8 + z_N \geq 1, \qquad \text{as the constaint for } R_0,$$
$$z_2 + z_3 + z_7 \geq 1, \qquad \text{as the constaint for } R_1,$$
$$\vdots \qquad\qquad\qquad\qquad \vdots$$
$$z_1 + z_3 + z_4 + z_9 \geq 1, \qquad \text{as the constaint for } R_{|\mathcal{R}|-1}.$$

Applications. We applied the new reduction algorithm to a system of inequalities generated with SAGE Math against various S-boxes. The results are shown in Table 3. Compared to the previous reduction algorithm based on the greedy algorithm, a smaller number of inequalities can be achieved for most of the applications. As a proof of context, the system of 21 inequalities to describe the DDT of PRESENT and LED is listed below.

```
- 1 x3 - 1 x2 + 0 x1 - 1 x0 - 1 y3 + 0 y2 + 1 y1 + 0 y0 + 3  >= 0,
- 1 x3 + 0 x2 - 1 x1 - 1 x0 + 1 y3 + 0 y2 - 1 y1 + 0 y0 + 3  >= 0,
  0 x3 - 2 x2 - 2 x1 - 2 x0 - 1 y3 + 2 y2 - 1 y1 - 1 y0 + 7  >= 0,
- 3 x3 + 2 x2 - 2 x1 - 1 x0 + 1 y3 - 2 y2 - 2 y1 - 1 y0 + 8  >= 0,
  1 x3 - 1 x2 + 1 x1 + 2 x0 - 2 y3 - 2 y2 + 1 y1 - 2 y0 + 5  >= 0,
  0 x3 - 1 x2 - 1 x1 + 1 x0 - 1 y3 + 0 y2 - 1 y1 + 1 y0 + 3  >= 0,
  2 x3 + 3 x2 - 2 x1 - 4 x0 - 4 y3 - 4 y2 - 1 y1 + 1 y0 + 11 >= 0,
  2 x3 - 1 x2 + 2 x1 + 2 x0 + 2 y3 + 3 y2 - 1 y1 - 1 y0 + 0  >= 0,
- 2 x3 + 1 x2 + 1 x1 + 3 x0 + 1 y3 - 1 y2 + 1 y1 + 2 y0 + 0  >= 0,
- 1 x3 + 1 x2 + 1 x1 - 1 x0 + 0 y3 + 0 y2 + 0 y1 - 1 y0 + 2  >= 0,
  0 x3 + 2 x2 - 2 x1 + 1 x0 - 1 y3 - 1 y2 - 2 y1 - 2 y0 + 6  >= 0,
  2 x3 + 3 x2 + 3 x1 + 2 x0 + 1 y3 - 4 y2 + 1 y1 + 1 y0 + 0  >= 0,
  1 x3 + 2 x2 + 2 x1 + 0 x0 - 1 y3 + 1 y2 - 1 y1 + 1 y0 + 0  >= 0,
  0 x3 - 2 x2 - 2 x1 + 3 x0 + 4 y3 + 1 y2 + 4 y1 + 1 y0 + 0  >= 0,
  2 x3 + 2 x2 - 1 x1 + 2 x0 - 1 y3 + 3 y2 + 2 y1 - 1 y0 + 0  >= 0,
  1 x3 + 3 x2 - 2 x1 - 2 x0 + 3 y3 + 4 y2 + 1 y1 + 4 y0 + 0  >= 0,
  1 x3 - 3 x2 - 2 x1 - 2 x0 + 3 y3 - 4 y2 + 1 y1 - 3 y0 + 10 >= 0,
- 1 x3 + 3 x2 + 3 x1 - 1 x0 + 2 y3 + 2 y2 + 2 y1 - 1 y0 + 0  >= 0,
  2 x3 - 2 x2 + 3 x1 - 4 x0 - 1 y3 - 4 y2 - 4 y1 + 1 y0 + 11 >= 0,
  1 x3 - 2 x2 + 3 x1 - 2 x0 + 1 y3 + 4 y2 + 3 y1 + 4 y0 + 0  >= 0,
- 2 x3 - 1 x2 - 1 x1 + 2 x0 - 2 y3 + 0 y2 - 2 y1 - 1 y0 + 7  >= 0.
```

Extension. With the new reduction algorithm, the users can specify the number of inequalities generated in the system by adding the following changes.

1. Add a constraint $\sum_{i=1}^{N} z_i = N_t$ where N_t is the number of inequalities to be included.
2. Leave the objective function empty.

5   Experiments

The new algorithm in Sect. 4 enables us to choose the number of inequalities in the system. In this section, we run the experiments which solve the fixed MILP model by choosing various numbers of inequalities in order to check how the number of inequalities is related to the runtime of the entire MILP. Sasaki and Todo presented how to model differential-trail search for LILLIPUT [23] in details. We determined to adopt their model to minimize inaccuracy. Section 5.1 briefly explains the specification of LILLIPUT. Then in Sect. 5.2 we report the experimental results.

5.1 LILLIPUT Specification

Block cipher LILLIPUT [15] was designed by Berger et al. in 2015, and adopts 16-branch extended generalized Feistel network [24] with the block-shuffle mechanism [25]. The block size and the key size are 64 bits and 80 bits, respectively. The state consists of 16 branches of size 4 bits. 64-bit plaintext is first loaded to sixteen nibbles $X_{15}, X_{14} \ldots, X_0$, and those are updated by the round function 30 times. The round function is illustrated in Fig. 3.

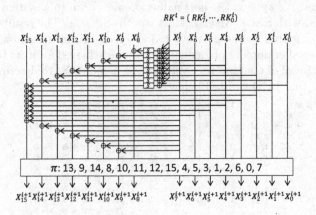

Fig. 3. Round function of LILLIPUT

At first, eight nibbles of round key are xored to each of eight nibbles in the right half of the state, and the results are xored to the left half of the state. Let RK_i^j and X_i^j be the i-th nibble of the j-th round key RK^j and j-th round state X^j, respectively. Then, the nonlinear layer can be defined as $X_{8+i}^j \leftarrow X_{8+i}^j \oplus S(X_{7-i}^j \oplus RK_i^j)$, $i = 0, 1, \ldots, 7$, where $S(\cdot)$ is a 4-bit to 4-bit S-box defined in Table 1.

After the non-linear update, some linear update applying xor between branches is performed, which is defined as follows.

$$X_{15}^j \leftarrow X_{15}^j \oplus X_7^j \oplus X_6^j \oplus X_5^j \oplus X_4^j \oplus X_3^j \oplus X_2^j \oplus X_1^j,$$
$$X_{15-i}^j \leftarrow X_{15-i}^j \oplus X_7^j \text{ for } i = 1, 2, \ldots, 6.$$

Finally, nibble positions are permuted according to π defined as

$$(13, 9, 14, 8, 10, 11, 12, 15, 4, 5, 3, 1, 2, 6, 0, 7)$$
$$\leftarrow \pi(0, 1, 2, 3, 4, 5, 6, 7, 8, 9, 10, 11, 12, 13, 14, 15).$$

Table 5. Runtime of differential-trail search for 5-round and 6-round LILLIPUT.

Time for 5 rounds

#inequalities	23	33	43	53	63	73	83	93
Time(sec)	75.68	106.30	19.95	28.33	14.77	24.77	20.45	18.92
#inequalities	103	113	123	133	143	153	163	173
Time(sec)	26.15	18.70	15.97	20.45	29.08	54.72	20.48	21.53
#inequalities	183	193	203	213	223	233	243	253
Time(sec)	20.51	68.53	28.35	22.96	24.38	20.55	27.33	28.91
#inequalities	263	273	283	293	303	313	323	
Time(sec)	28.66	30.50	25.96	29.14	33.46	33.49	34.41	

Time for 6 rounds

#inequalities	23	33	43	53	63	73	83	93
Time(sec)	14068.13	1125.83	1333.97	2941.26	1351.91	2640.21	2424.82	1092.05
#inequalities	103	113	123	133	143	153	163	173
Time(sec)	1205.98	1411.40	1495.90	1330.74	1395.17	1376.73	1538.08	1805.86
#inequalities	183	193	203	213	223	233	243	253
Time(sec)	1841.54	1272.60	1893.98	2402.03	3938.68	5401.42	3110.85	3060.52
#inequalities	263	273	283	293	303	313	323	
Time(sec)	2203.71	2060.90	3565.50	3443.10	3558.77	4784.93	5104.19	

5.2 Runtime of Differential Trail Search for LILLIPUT

We first generated the system of inequalities to model the DDT of LILLIPUT's
S-box with SAGE Math, which returned 324 inequalities. We then ran the reduction algorithm described in Sect. 4, which revealed that all valid propagations
can be described with 23 inequalities in minimum as summarized in Table 3.

The goal of this section is performing experiments to detect the relationship
between the number of inequalities of the single DDT and the runtime of the
entire differential-trail search problem. We tested 5-round differential-trail search
and 6-round differential-trail search with $23 + 10i$ inequalities per S-box, where
we change i within the range of $0 \leq i \leq 30$. The results are summarized in
Table 5 and in Fig. 4.

The results clearly show that the runtime is significantly slow when the number of inequalities for each S-box is almost minimum. It is even slower than the
case that no reduction is performed to the system of inequalities.

This phenomenon can be intuitively explained as follows. The MILP solver
itself optimizes the system by considering the entire system instead of focusing
on each S-box. Thus, leaving some room for the solver to optimize can result
in the minimum runtime for the entire system. However, including too many
inequalities also consumes time to optimize it. Thus, as the number of inequalities
becomes significantly larger, the runtime gets gradually bigger.

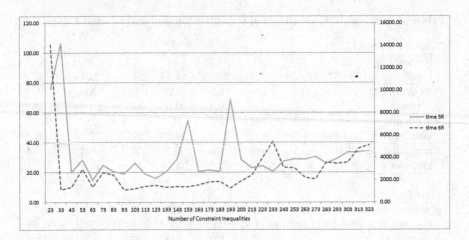

Fig. 4. Graphical representation of the experimental results. The left vertical axis is for 5 rounds and the right vertical axis is for 6 rounds.

6 Application to Division-Trail Search

Division property was invented by Todo [26] as a new tool to evaluate the property used in integral cryptanalysis in a more general form. The concept was later extended to bit-based division property by Todo and Morii [27], which needs to evaluate the S-box property with the division property defined in each bit.

To evaluate the S-box, it first obtains the algebraic normal form of the S-box, for example is described as

$$y_0 = 1 + x_0 + x_1 + x_0 x_1 + x_2 + x_3,$$
$$y_1 = x_0 + x_0 x_1 + x_2 + x_0 x_2 + x_3,$$
$$y_2 = x_1 + x_2 + x_0 x_3 + x_1 x_3 + x_1 x_2 x_3,$$
$$y_3 = x_0 + x_1 x_3 + x_0 x_2 x_3,$$

and the propagation of the division property is summarized in a table, which is a counterpart of DDT in differential cryptanalysis. For example, the table of the division property for the above S-box is shown in Table 6. In this table, u and v are the input and output division property, respectively, and the propagation from u to v labeled x is possible. Otherwise, the propagation is impossible.

Xiang et al. found that the division trails can be searched with MILP [28] and the search was extended to bit-based division property by Sun et al. [29]. Then, modeling the above table, namely excluding the impossible patterns in the above table is necessary. Due to the same format as modeling DDT, it can be done by generating a large matrix with SAGE Math or logical computation model, and then applying the reduction algorithm. Here, our new reduction algorithm can be applied similarly as modeling DDT.

Table 6. Possible propagations of the division property for an S-box.

$\frac{v}{u}$	0	1	2	4	8	3	5	9	6	A	C	7	B	D	E	F
0	x	x	x	x	x	x	x	x	x	x	x	x	x	x	x	x
1		x	x	x	x	x	x	x	x	x	x	x	x	x	x	x
2		x	x	x	x	x	x	x	x	x	x	x	x	x	x	x
4		x	x	x	x	x	x	x	x	x	x	x	x	x	x	x
8		x	x	x	x	x	x	x	x	x	x	x	x	x	x	x
3		x	x		x	x	x	x	x	x	x	x	x	x	x	x
5		x		x	x	x	x	x	x	x	x	x	x	x	x	x
9			x	x	x	x	x	x	x	x	x	x	x	x	x	x
6			x		x	x	x	x	x	x	x	x	x	x	x	x
A			x	x	x	x	x	x	x	x	x	x	x	x	x	x
C			x	x	x	x	x	x	x	x	x	x	x	x	x	x
7							x					x	x	x	x	x
B								x				x	x	x	x	x
D			x	x	x	x					x	x	x	x	x	x
E			x				x	x	x	x	x	x	x	x	x	x
F																x

7 Concluding Remarks

In this paper, we revisited the reduction algorithm to model DDT by using MILP. Compared to previous one with the greedy algorithm, our method which is based on another small MILP problem can ensure the minimal number of inequalities to model DDT. It also enables users to choose the number of inequalities included in the system.

We then solved a 5-round and 6-round differential bound search for LILLIPUT with various numbers of inequalities per S-box. The results show that, on the contrary to the previous belief, minimizing the number of inequalities increases the runtime of the entire problem significantly.

We also discussed that the same reduction algorithm can be used to model the possible propagation patterns of the division property for S-box.

Many researches have been done on the automated search with MILP. Because most of the papers do not explain how many inequalities are used to model DDT, it is hard to distinguish which research can be improved. However, we believe that the general knowledge provided by this paper helps optimizing the speed of the differential search, hence helps future implementors.

References

1. Biham, E., Shamir, A.: Differential Cryptanalysis of the Data Encryption Standard. Springer, New York (1993). doi:10.1007/978-1-4613-9314-6
2. Mouha, N., Wang, Q., Gu, D., Preneel, B.: Differential and linear cryptanalysis using mixed-integer linear programming. In: Wu, C.-K., Yung, M., Lin, D. (eds.) Inscrypt 2011. LNCS, vol. 7537, pp. 57–76. Springer, Heidelberg (2012). doi:10.1007/978-3-642-34704-7_5

3. Bogdanov, A., Knudsen, L.R., Leander, G., Paar, C., Poschmann, A., Robshaw, M.J.B., Seurin, Y., Vikkelsoe, C.: PRESENT: an ultra-lightweight block cipher. In: Paillier, P., Verbauwhede, I. (eds.) CHES 2007. LNCS, vol. 4727, pp. 450–466. Springer, Heidelberg (2007). doi:10.1007/978-3-540-74735-2_31

4. Grosso, V., Leurent, G., Standaert, F.-X., Varıcı, K.: LS-designs: bitslice encryption for efficient masked software implementations. In: Cid, C., Rechberger, C. (eds.) FSE 2014. LNCS, vol. 8540, pp. 18–37. Springer, Heidelberg (2015). doi:10.1007/978-3-662-46706-0_2

5. Sun, S., Hu, L., Wang, P., Qiao, K., Ma, X., Song, L.: Automatic security evaluation and (Related-key) differential characteristic search: application to SIMON, PRESENT, LBlock, DES(L) and other bit-oriented block ciphers. In: Sarkar, P., Iwata, T. (eds.) ASIACRYPT 2014. LNCS, vol. 8873, pp. 158–178. Springer, Heidelberg (2014). doi:10.1007/978-3-662-45611-8_9

6. Sun, S., Hu, L., Wang, M., Wang, P., Qiao, K., Ma, X., Shi, D., Song, L., Fu, K.: Towards finding the best characteristics of some bit-oriented block ciphers and automatic enumeration of (related-key) differential and linear characteristics with predefined properties. Cryptology ePrint Archive, Report 2014/747 (2014)

7. Gong, Z., Nikova, S., Law, Y.W.: KLEIN: a new family of lightweight block ciphers. In: Juels, A., Paar, C. (eds.) RFIDSec 2011. LNCS, vol. 7055, pp. 1–18. Springer, Heidelberg (2012). doi:10.1007/978-3-642-25286-0_1

8. Shibutani, K., Isobe, T., Hiwatari, H., Mitsuda, A., Akishita, T., Shirai, T.: Piccolo: an ultra-lightweight blockcipher. In: Preneel, B., Takagi, T. (eds.) CHES 2011. LNCS, vol. 6917, pp. 342–357. Springer, Heidelberg (2011). doi:10.1007/978-3-642-23951-9_23

9. Suzaki, T., Minematsu, K., Morioka, S., Kobayashi, E.: TWINE: a lightweight block cipher for multiple platforms. In: Knudsen, L.R., Wu, H. (eds.) SAC 2012. LNCS, vol. 7707, pp. 339–354. Springer, Heidelberg (2013). doi:10.1007/978-3-642-35999-6_22

10. Borghoff, J., et al.: PRINCE – a low-latency block cipher for pervasive computing applications. In: Wang, X., Sako, K. (eds.) ASIACRYPT 2012. LNCS, vol. 7658, pp. 208–225. Springer, Heidelberg (2012). doi:10.1007/978-3-642-34961-4_14

11. Izadi, M., Sadeghiyan, B., Sadeghian, S.S., Khanooki, H.A.: MIBS: a new lightweight block cipher. In: Garay, J.A., Miyaji, A., Otsuka, A. (eds.) CANS 2009. LNCS, vol. 5888, pp. 334–348. Springer, Heidelberg (2009). doi:10.1007/978-3-642-10433-6_22

12. Guo, J., Peyrin, T., Poschmann, A., Robshaw, M.: The LED block cipher. In: Preneel, B., Takagi, T. (eds.) CHES 2011. LNCS, vol. 6917, pp. 326–341. Springer, Heidelberg (2011). doi:10.1007/978-3-642-23951-9_22

13. Wu, W., Zhang, L.: LBlock: a lightweight block cipher. In: Lopez, J., Tsudik, G. (eds.) ACNS 2011. LNCS, vol. 6715, pp. 327–344. Springer, Heidelberg (2011). doi:10.1007/978-3-642-21554-4_19

14. Biham, E., Anderson, R., Knudsen, L.: Serpent: a new block cipher proposal. In: Vaudenay, S. (ed.) FSE 1998. LNCS, vol. 1372, pp. 222–238. Springer, Heidelberg (1998). doi:10.1007/3-540-69710-1_15

15. Berger, T.P., Francq, J., Minier, M., Thomas, G.: Extended generalized Feistel networks using matrix representation to propose a new lightweight block cipher: lilliput. IEEE Trans. Comput. **65**, 2074–2089 (2015)

16. Banik, S., Bogdanov, A., Isobe, T., Shibutani, K., Hiwatari, H., Akishita, T., Regazzoni, F.: Midori: a block cipher for low energy. In: Iwata, T., Cheon, J.H. (eds.) ASIACRYPT 2015. LNCS, vol. 9453, pp. 411–436. Springer, Heidelberg (2015). doi:10.1007/978-3-662-48800-3_17

17. Sasaki, Y., Todo, Y., Aoki, K., Naito, Y., Sugawara, T., Murakami, Y., Matsui, M.: Minalpher v1.1. Submitted to CAESAR (2015)
18. Zhang, W., Bao, Z., Lin, D., Rijmen, V., Yang, B., Verbauwhede, I.: RECTAN-GLE: a bit-slice lightweight block cipher suitable for multiple platforms. Cryptology ePrint Archive, Report 2014/084 (2014)
19. Beierle, C., et al.: The SKINNY family of block ciphers and its low-latency variant MANTIS. In: Robshaw, M., Katz, J. (eds.) CRYPTO 2016. LNCS, vol. 9815, pp. 123–153. Springer, Heidelberg (2016). doi:10.1007/978-3-662-53008-5_5
20. Gurobi Optimization Inc.: Gurobi Optimizer 6.5. Official webpage (2015). http://www.gurobi.com/
21. Gamrath, G., Fischer, T., Gally, T., Gleixner, A.M., Hendel, G., Koch, T., Maher, S.J., Miltenberger, M., Müller, B., Pfetsch, M.E., Puchert, C., Rehfeldt, D., Schenker, S., Schwarz, R., Serrano, F., Shinano, Y., Vigerske, S., Weninger, D., Winkler, M., Witt, J.T., Witzig, J.: The SCIP Optimization Suite 3.2. Technical report 15–60, ZIB, Takustr. 7, 14195 Berlin (2016)
22. Ilog, I.: IBM ILOG CPLEX Optimization Studio V12.7.0 documentation. Official webpage (2016). https://www-01.ibm.com/software/websphere/products/optimization/cplex-studio-community-edition/
23. Sasaki, Y., Todo, Y.: New differential bounds and division property of shape LIL-LIPUT: block cipher with extended generalized Feistel network. In: Avanzi, R., Heys, H. (eds.) SAC 2016. LNCS, Springer, (to appear 2016). Pre-proceedings version was distributed at the workshop
24. Berger, T.P., Minier, M., Thomas, G.: Extended generalized feistel networks using matrix representation. In: Lange, T., Lauter, K., Lisoněk, P. (eds.) SAC 2013. LNCS, vol. 8282, pp. 289–305. Springer, Heidelberg (2014). doi:10.1007/978-3-662-43414-7_15
25. Suzaki, T., Minematsu, K.: Improving the generalized feistel. In: Hong, S., Iwata, T. (eds.) FSE 2010. LNCS, vol. 6147, pp. 19–39. Springer, Heidelberg (2010). doi:10.1007/978-3-642-13858-4_2
26. Todo, Y.: Structural evaluation by generalized integral property. In: Oswald, E., Fischlin, M. (eds.) EUROCRYPT 2015. LNCS, vol. 9056, pp. 287–314. Springer, Heidelberg (2015). doi:10.1007/978-3-662-46800-5_12
27. Todo, Y., Morii, M.: Bit-based division property and application to SIMON family. In: Peyrin, T. (ed.) FSE 2016. LNCS, vol. 9783, pp. 357–377. Springer, Heidelberg (2016). doi:10.1007/978-3-662-52993-5_18
28. Xiang, Z., Zhang, W., Bao, Z., Lin, D.: Applying MILP method to searching integral distinguishers based on division property for 6 lightweight block ciphers. In: Cheon, J.H., Takagi, T. (eds.) ASIACRYPT 2016. LNCS, vol. 10031, pp. 648–678. Springer, Heidelberg (2016). doi:10.1007/978-3-662-53887-6_24
29. Sun, L., Wang, W., Liu, R., Wang, M.: MILP-Aided Bit-Based Division Property for ARX-Based Block Cipher. Cryptology ePrint Archive, Report 2016/1101 (2016)

Secretly Embedding Trapdoors into Contract Signing Protocols

Diana Maimuţ[1(✉)] and George Teşeleanu[1,2]

[1] Advanced Technologies Institute, 10 Dinu Vintilă, Bucharest, Romania
ati@dcti.ro
[2] Department of Computer Science, "Al.I.Cuza" University of Iaşi,
700506 Iaşi, Romania
george.teseleanu@info.uaic.ro

Abstract. Contract signing protocols have been proposed and analyzed for more than three decades now. One of the main problems that appeared while studying such schemes is the impossibility of achieving both fairness and guaranteed output delivery. As workarounds, cryptographers have put forth three main categories of contract signing schemes: gradual release, optimistic and concurrent or legally fair schemes. Concurrent signature schemes or legally fair protocols do not rely on trusted arbitrators and, thus, may seem more attractive for users. Boosting user trust in such manner, an attacker may cleverly come up with specific applications. Thus, our work focuses on embedding trapdoors into contract signing protocols. In particular, we describe and analyze various SETUP (Secretly Embedded Trapdoor with Universal Protection) mechanisms which can be injected in concurrent signature schemes and legally fair protocols without keystones.

1 Introduction

Contract signing protocols have been proposed and extensively studied in the past. During the analysis of such schemes, the impossibility of achieving both fairness and guaranteed output delivery became a central problem for researchers. Trying to solve the aforementioned issue, cryptographers have developed various contract signing schemes which can be categorized having in mind three different design types: ① *gradual release* [12,14,15,18], ② *optimistic* [2,5,17] and ③ *concurrent* [6] or *legally fair* [10] models. Concurrent signatures or legally fair protocols do not rely on trusted third parties. Also, concurrent signature models do not require too much interaction between users as compared to older paradigms like gradual release or optimistic models. Such features may seem much more attractive for users. Building upon user trust in the case of fair contract signing protocols, a (powerful) adversary may cleverly construct attack scenarios.

Digital signature schemes naturally arose as the central ingredient of modern contract signing protocols. The use of digital signatures as a channel to convey information (subliminal channel) was first introduced and studied by Simmons

© Springer International Publishing AG 2017
P. Farshim and E. Simion (Eds.): SecITC 2017, LNCS 10543, pp. 166–186, 2017.
https://doi.org/10.1007/978-3-319-69284-5_12

[21,22]. Another step was taken by Young and Yung [23–27], who combined subliminal channels and public key cryptography in order to leak a user's private key (SETUP attacks). The two authors work in a black-box environment[1], pointing out that other scenarios exist. Such attacks may be considered if the manufacturer of the device is an accomplice, in the sense that he implements the mechanisms to recover the keys.

A SETUP attack of the previously mentioned form is likely to be applied in the case of auctions. To provide the reader with a possible scenario, we further assume that participants receive signing tokens from an auctioneer and they do not communicate using additional channels. The participants' bids are acknowledged by the auctioneer's co-signature. In this context, the auctioneer is able to leak lists containing fake bids for the competing participants. The value of the bids is, thus, maliciously raised.

Our work focuses on embedding trapdoors into contract signing protocols. In particular, we describe and analyze two main SETUP mechanisms which can be injected in the concurrent signature scheme presented in [6] and the legally fair protocol (without keystones) introduced in [10].

Structure of the Paper. We introduce notations, definitions and protocols used throughout the paper in Sect. 2. In Sects. 3 and 4 we present two main SETUP mechanisms which can be injected into concurrent or legally fair signature schemes and analyze their security in the standard model and, respectively, Random Oracle Model (ROM) [3]. We conclude in Sect. 5. We recall additional security models and Schnorr signatures in Appendix A and provide supplementary SETUP mechanisms in Appendices B and C.

2   Preliminaries

Notations. Let S be a finite set. We denote by $x \xleftarrow{\$} S$ the operation of picking an element uniformly from S.

$x\|y$ represents the string obtained by concatenating y to x.

If and only if is further referred to as *iff*.

Unless otherwise specified, $\mathbb{G}$ is a cyclic group of order q, where q is a large prime number. Also, we denote by g a generator of $\mathbb{G}$.

x_i and y_i represent the private and public keys associated with user i: x_i is considered to be randomly chosen from $\mathbb{Z}_q^*$ and $y_i = g^{x_i}$.

The action of choosing a random element from an entropy smoothing[2] (ES) family $\mathcal{H}$ is further referred to as "H is ES".

We denote by PPT algorithm a probabilistic polynomial-time algorithm.

---

[1] *e.g.* tamper proof devices.

[2] We refer the reader to Appendix A for a definition of the concept.

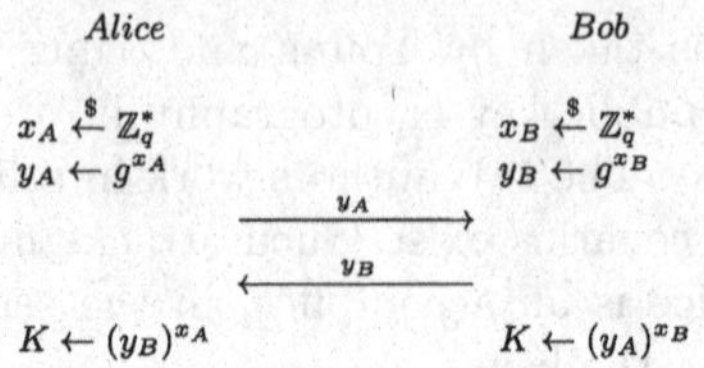

Fig. 1. The Diffie-Hellman key exchange protocol.

2.1 Security Assumptions

Definition 1 (Discrete Logarithm Problem - DLP). *Let $\mathbb{G}$ be a cyclic group of order n and g a generator $\mathbb{G}$. Given $g, h \overset{\$}{\leftarrow} \mathbb{G}$, find a such that $h = g^a$.*

The number a is called the discrete logarithm of h to the base g and is denoted by $\log_g h$.

Remark 1. Two users A and B can choose a DLP based protocol in order to compute a common secret key K. We describe the Diffie-Hellman (DH) key exchange [7] in Fig. 1.

Definition 2 (Computational Diffie-Hellman - CDH and List Computational Diffie-Hellman of Order 2 - LCDH2). *Let $\mathbb{G}$ be a cyclic group of order n, g a generator $\mathbb{G}$ and let A be a PPT algorithm that returns either an element (CDH) or a list of elements (LCDH2) from $\mathbb{G}$. We define the advantages*

$$ADV_{\mathbb{G},g}^{\mathrm{CDH}}(A) = Pr[A(g^x, g^y) = g^{xy} | x, y \overset{\$}{\leftarrow} \mathbb{Z}_n^*]$$

$$ADV_{\mathbb{G},g}^{\mathrm{LCDH2}}(A) = Pr[g^{xy} \text{ or } g^{xz} \in A(g^x, g^y, g^z) | x, y, z \overset{\$}{\leftarrow} \mathbb{Z}_n^*].$$

If $ADV_{\mathbb{G},g}^{\mathrm{CDH}}(A)$ or $ADV_{\mathbb{G},g}^{\mathrm{LCDH2}}(A)$ is negligible for any PPT algorithm A, we say that the Computational Diffie-Hellman problem or List Computational Diffie-Hellman problem of Order 2 is hard in $\mathbb{G}$.

Remark 2. A similar with LCDH2 concept was introduced in [20] and proven to be equivalent with CDH. Tweaking the proof from [20], we obtain that for an efficient PPT LCDH2 adversary A there exist an efficient PPT algorithm B such that

$$ADV_{\mathbb{G},g}^{\mathrm{LCDH2}}(A) \le 2ADV_{\mathbb{G},g}^{\mathrm{CDH}}(B). \tag{1}$$

It is easy to see that if the CDH assumption doesn't hold, then the LCDH2 assumption doesn't hold. If the LCDH2 assumption doesn't hold, then there exist a PPT algorithm A that has non-negligible LCDH2 advantage. We will use A to build an algorithm B that has non-negligible CDH advantage for (g^x, g^y) or (g^x, g^z). Algorithm B simply runs A and then outputs two random elements from the list returned by A. Thus we obtain the loose reduction (1).

Definition 3 (Decisional Diffie-Hellman - DDH). *Let $\mathbb{G}$ be a cyclic group of order n, g a generator $\mathbb{G}$ and let A be a PPT algorithm. We define the advantage*

$$ADV_{\mathbb{G},g}^{\mathrm{DDH}}(A) = \Big| Pr[A(g^x, g^y, g^z) = 1 | x, y \xleftarrow{\$} \mathbb{Z}_n^*, z \leftarrow xy]$$

$$-Pr[A(g^x, g^y, g^z) = 1 | x, y, z \xleftarrow{\$} \mathbb{Z}_n^*] \Big|.$$

If $ADV_{\mathbb{G},g}^{\mathrm{DDH}}(A)$ is negligible for any PPT algorithm A, we say that the Decisional Diffie-Hellman problem is hard in $\mathbb{G}$.

2.2 Security Models

Definition 4 (Pseudorandom Function - PRF). *A function $F : \{0,1\}^n \times \{0,1\}^s \to \{0,1\}^m$ is a (t,q)-PRF if:*

- *Given a key $K \in \{0,1\}^s$ and an input $X \in \{0,1\}^n$ there is an efficient algorithm to compute $F_K(X) = F(X, K)$.*
- *For any t-time oracle algorithm A, the PRF-advantage of A, defined as*

$$ADV_F^{\mathrm{PRF}}(A) = \Big| Pr[A^{F_K(\cdot)} = 1 | K \xleftarrow{\$} \{0,1\}^s] - Pr[A^{F(\cdot)} = 1 | F \xleftarrow{\$} \mathcal{F}] \Big|$$

is negligible for any PPT algorithm A, where $\mathcal{F} = \{F : \{0,1\}^n \to \{0,1\}^m\}$ and A makes at most q queries to the oracle.

Definition 5 (Secretly Embedded Trapdoor with Universal Protection - SETUP). *A Secretly Embedded Trapdoor with Universal Protection (SETUP) is an algorithm that can be inserted in a system such that it leaks encrypted private key information to an attacker through the system's outputs. The leakage is achieved through a public key exchange protocol between an unsuspecting victim and the attacker.*

Definition 6 (SETUP indistinguishability - IND-SETUP). *Let C_0 be a black-box system that uses a pair of keys (pk, sk), where pk is the public key and sk the corresponding secret key. Let pk_S be the public key of an attacker as defined in Definition 5. Let $\mathcal{KE}$ be a public key exchange protocol that takes as input pk and pk_S. We consider C_1 an altered version of C_0 that contains a SETUP mechanism based on $\mathcal{KE}$. Let A be a PPT algorithm. We define the advantage*

$$ADV_{\mathcal{KE},C_0,C_1}^{\mathrm{IND\text{-}SETUP}}(A) = \Big| Pr[A^{C_1(sk,\cdot)}(pk, pk_S) = 1]$$

$$-Pr[A^{C_0(sk,\cdot)}(pk, pk_S) = 1] \Big|.$$

If $ADV_{\mathcal{KE},C_0,C_1}^{\mathrm{IND\text{-}SETUP}}(A)$ is negligible for any PPT algorithm A, we say that C_0 and C_1 are polynomially indistinguishable.

2.3 Concurrent Signatures

In the case of classical contract signing protocols, users exchange complete signatures (*e.g.* [13]). Concurrent signature protocols [6,16] use "ambiguous" signatures that do not bind their author. An additional piece of information called the *keystone* can later be used to lift the ambiguity. Thus, when the keystone is revealed, signatures become simultaneously binding.

The standard algorithms corresponding to a concurrent signature are shortly described in Table 1.

Table 1. The algorithms of a concurrent signature.

Setup(ℓ)	On input a security parameter ℓ, this algorithm outputs the private and public keys (x_i, y_i) of all participants and the public parameters $pp = (\mathcal{M}, \mathcal{K}, \mathcal{F}, KeyGen)$, where $KeyGen : \mathcal{K} \to \mathcal{F}$ is a selected function.
aSign(y_i, y_j, x_i, e_2, m)	On input the public keys y_i and y_j, the private key x_i corresponding to y_i, an element $e_2 \in \mathcal{F}$ and a message $m \in \mathcal{M}$, this algorithm outputs an "ambiguous signature" $\sigma = \langle s, e_1, e_2 \rangle$, where $s \in \mathcal{S}$ and $e_1, e_2 \in \mathcal{F}$.
aVerify(σ, y_i, y_j, m)	On input an ambiguous signature $\sigma = \langle s, e_1, e_2 \rangle$, public keys y_i, y_j and a message m this algorithm outputs a boolean value satisfying $$aVerify\left(\sigma', y_j, y_i, m\right) = aVerify\left(\sigma, y_i, y_j, m\right),$$ where $\sigma' = \langle s, e_2, e_1 \rangle$.
Verify(k, σ, y_i, y_j, m)	On input $k \in \mathcal{K}$, $\sigma = \langle s, e_2, e_1 \rangle$, public keys y_i, y_j and message m, this algorithm checks whether $KeyGen(k) = e_2$ and outputs False if not; otherwise it outputs the result of aVerify(σ, y_i, y_j, m).

Concurrent signatures are used by two parties *Alice* and *Bob* as depicted in Fig. 2.

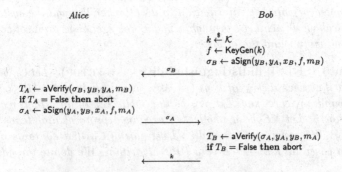

Fig. 2. The concurrent signature of messages m_A and m_B.

At the end of this protocol, both $\langle k, \sigma_A \rangle$ and $\langle k, \sigma_B \rangle$ are binding, and accepted by the Verify algorithm.

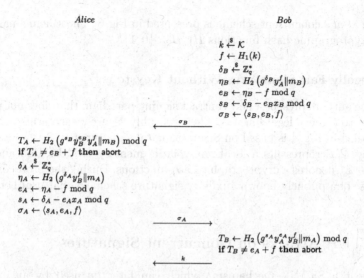

Fig. 3. Chen *et al.* concurrent signature.

A Concrete Construction. To mount our SETUP attacks, we further use a concrete concurrent signature, more precisely the protocol presented in [6]. The security of this protocol can be proven in the ROM, assuming the hardness of computing discrete logarithms in a group $\mathbb{G}$.

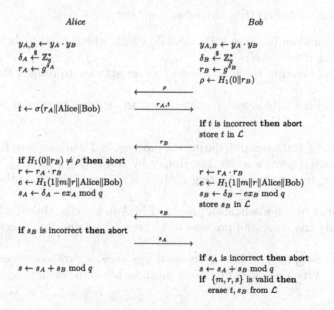

Fig. 4. The legally fair signature (without keystones) of message m.

Chen *et al*'s concurrent scheme is presented in Fig. 3. The scheme makes use of two cryptographic hash functions $H_1, H_2 : \{0,1\}^* \rightarrow \mathbb{Z}_q^*$.

2.4 Legally Fair Signatures Without Keystones

In [10] the authors present a new contract signing paradigm that does not require keystones to achieve legal fairness. Their provably secure co-signature construction recalled in Fig. 4 is based on Schnorr digital signatures[3].

In Fig. 4, $\mathcal{L}$ represents a local non-volatile memory used by Bob and $H_1 : \{0,1\}^* \rightarrow \mathbb{Z}_q^*$ denotes a cryptographic hash functions. During the protocol, Alice makes use of a publicly known auxiliary signature scheme σ that uses her secret key x_A.

3 SETUP Attacks on Concurrent Signatures

We present a SETUP mechanism[4] which can later be used by an external attacker Eve to recover either Alice's or Bob's secret key. To implement her attack, Eve needs a valid pair of (private and public) keys $(x_E, y_E = g^{x_E})$. The public key y_E is stored in a volatile memory on the victim's device. We further assume that Eve has access to the data transmitted during the protocol.

Changes required by the SETUP mechanisms will further be underlined using red colored text within Fig. 5.

Description. The SETUP mechanism requires:

– a pseudorandom function $\mathbb{PRF} : K \times \mathbb{Z}_q^* \rightarrow \mathbb{Z}_q^*$, where K is the key space;
– a function $H : \mathbb{G} \rightarrow K$;
– a protocol needs to reach breakpoint ① for an attacker to recover Bob's secret key;
– a protocol needs to reach breakpoint ② for an attacker to recover both secret key.

The value f is transmitted during the protocol and is available to Eve. Hence, she can recover user i's secret key simply by computing $\mathbb{PRF}(H(y_i^{x_E}), f)$ and extracting x_i from s_i by calculating $e_i^{-1}(\delta_i - s_i)$, where i denotes either Alice or Bob.

Compared to the mechanism presented in Appendix B, this SETUP attack requires only one successful protocol to recover Alice's and Bob's secret key.

Malicious Co-signers. If Eve is replaced by Alice, a protocol needs to reach breakpoint ①. When replaced by Bob, a protocol needs to reach breakpoint ②.

---

[3] Recalled in Appendix A.
[4] Another mechanism (detailed in Appendix B) naturally arises.

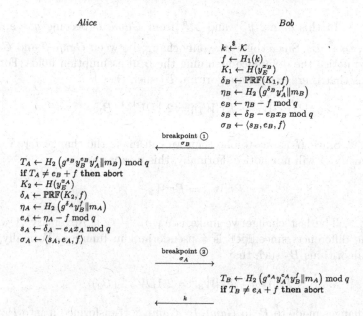

$$Alice \qquad\qquad\qquad\qquad\qquad\qquad Bob$$

$$k \xleftarrow{\$} \mathcal{K}$$
$$f \leftarrow H_1(k)$$
$$K_1 \leftarrow H(y_E^{x_B})$$
$$\delta_B \leftarrow \mathrm{PRF}(K_1, f)$$
$$\eta_B \leftarrow H_2\left(g^{\delta_B} y_A^f \| m_B\right)$$
$$e_B \leftarrow \eta_B - f \bmod q$$
$$s_B \leftarrow \delta_B - e_B x_B \bmod q$$
$$\sigma_B \leftarrow \langle s_B, e_B, f \rangle$$

breakpoint ①
$$\xleftarrow{\quad \sigma_B \quad}$$

$$T_A \leftarrow H_2\left(g^{s_B} y_B^{e_B} y_A^f \| m_B\right) \bmod q$$
if $T_A \neq e_B + f$ then abort
$$K_2 \leftarrow H(y_E^{x_A})$$
$$\delta_A \leftarrow \mathrm{PRF}(K_2, f)$$
$$\eta_A \leftarrow H_2\left(g^{\delta_A} y_B^f \| m_A\right)$$
$$e_A \leftarrow \eta_A - f \bmod q$$
$$s_A \leftarrow \delta_A - e_A x_A \bmod q$$
$$\sigma_A \leftarrow \langle s_A, e_A, f \rangle$$

breakpoint ②
$$\xrightarrow{\quad \sigma_A \quad}$$

$$T_B \leftarrow H_2\left(g^{s_A} y_A^{e_A} y_B^f \| m_A\right) \bmod q$$
if $T_B \neq e_A + f$ then abort

$$\xleftarrow{\quad k \quad}$$

Fig. 5. The Protocol presented in Fig. 3 with a SETUP mechanism. (Color figure online)

Security Analysis. We present the main security results, more precisely Theorems 1 and 2, and provide the reader with the necessary proofs.

When referring to the security analysis presented in the current section, Θ is considered an additional security parameter and refers to the maximal number of protocol iterations.

Theorem 1. *If* DDH *is hard in* $\mathbb{G}$ *and* H *is a one-to-one function[5], then the protocols presented in Figs. 3 and 5 are* IND-SETUP *in the standard model. Formally, let* A *be an efficient PPT* IND-SETUP *adversary. There exist two efficient PPT algorithms* B_1, B_2 *such that*

$$ADV_{\mathrm{DH},P_3,P_5}^{\mathrm{IND\text{-}SETUP}}(A) \leq 4ADV_{\mathbb{G},g}^{\mathrm{DDH}}(B_1) + 4ADV_{\mathrm{PRF}}^{\mathrm{PRF}}(B_2).$$

Proof. We denote the protocols presented in Figs. 3 and 5 by P_3 and P_5. Let A be an IND-SETUP adversary trying to distinguish between P_3 and P_5. We show that A's advantage is negligible. We construct the proof as a sequence of games in which all the required changes are applied to P_5. Let W_i be the event that A wins game i.

Game 0. The first game is identical to the IND-SETUP game[6]. Thus, we have

$$|2Pr[W_0] - 1| = ADV_{\mathrm{DH},P_3,P_5}^{\mathrm{IND\text{-}SETUP}}(A). \tag{2}$$

---

[5] A function for which every element of the range of the function corresponds to precisely one element of the domain.

[6] As in Definition 6.

Game 1. In this game, $y_E^{x_A}$ and $y_E^{x_B}$ from *Game 0* become g^{z_A} and g^{z_B}, where $z_A, z_B \xleftarrow{\$} \mathbb{Z}_q$. Since this is the only change between *Game 0* and *Game 1*, A will not notice the difference assuming the DDH assumption holds. Formally, this means that there exists an algorithm B_1 such that

$$|Pr[W_0] - Pr[W_1]| = 2ADV_{\mathbb{G},g}^{\mathrm{DDH}}(B_1). \tag{3}$$

Game 2. Since H is one-to-one then we can make the change $K_1, K_2 \xleftarrow{\$} \mathbb{Z}_q$ and adversary A will not notice. Formally, this means that

$$Pr[W_1] = Pr[W_2]. \tag{4}$$

Game 3. The last change we make is $\delta_A, \delta_B \xleftarrow{\$} \mathbb{Z}_q$. Adversary A will not notice the difference, since $\mathbb{PRF}$ is a pseudorandom function. Formally, there exist an algorithms B_2 such that

$$|Pr[W_2] - Pr[W_3]| = 2ADV_{\mathrm{PRF}}^{\mathrm{PRF}}(B_2). \tag{5}$$

The changes made to P_5 in *Game 1 - Game 3*, transformed it into P_3. Thus, we have

$$Pr[W_3] = 1/2. \tag{6}$$

Finally, the statement is proven by combining the equalities (2)–(6).   □

Remark 3. From Theorem 1, the maximum advantage an IND-SETUP adversary can obtain in the standard model is

$$ADV_{\mathrm{DH},P_3,P_5}^{\mathrm{IND\text{-}SETUP}}(A) \le 4\Theta ADV_{\mathbb{G},g}^{\mathrm{DDH}}(B_1) + 4\Theta ADV_{\mathrm{PRF}}^{\mathrm{PRF}}(B_2).$$

The advantage remains negligible if parameter Θ is polynomial.

Theorem 2. *If CDH is hard in $\mathbb{G}$ and H is a hash function, then the protocols presented in Figs. 3 and 5 are IND-SETUP in the ROM. Formally, let A be an efficient PPT IND-SETUP adversary. There exist two efficient PPT algorithms B_1, B_2 such that*

$$ADV_{\mathrm{DH},P_3,P_5}^{\mathrm{IND\text{-}SETUP}}(A) \le 4ADV_{\mathbb{G},g}^{\mathrm{CDH}}(B_1) + 4ADV_{\mathrm{PRF}}^{\mathrm{PRF}}(B_2).$$

Proof. We will use the same notations as in the proof for Theorem 1.

Game 0. The first game is identical to the IND-SETUP game[7]. Thus, we have

$$|2Pr[W_0] - 1| = ADV_{\mathrm{DH},P_3,P_5}^{\mathrm{IND\text{-}SETUP}}(A). \tag{7}$$

The challenger picks a random oracle $H : \mathbb{G} \to \mathbb{Z}_q^*$ at random from the set of all such functions. A can make a sequence of queries of the following type.

---

[7] see Footnote 6.

Hash oracle query[8]: A presents the challenger with $m \in \mathbb{G}$, who responds with $H(m)$.

Game 1. At the beginning of the game choose $K_1, K_2 \xleftarrow{\$} \mathbb{Z}_q^*$. The challenger's way to respond to queries becomes:

Hash oracle query[9]: A presents the challenger with $m \in \mathbb{G}$. The challenger responds with

- K_1, if $m = y_E^{x_A}$;
- K_2, if $m = y_E^{x_B}$;
- $H(m)$, otherwise.

Since we have replaced the values $y_E^{x_A}$ and $y_E^{x_B}$ throughout the game, we have

$$Pr[W_0] = Pr[W_1]. \tag{8}$$

Game 2. In this game, we revert to the original hash oracle query (*i.e* the challenger responds with $H(m)$ for all m). Let F be the event that the adversary makes a query with $m \leftarrow y_E^{x_A}$ or $m \leftarrow y_E^{x_B}$. *Game 1* and *Game 2* are identical until F occurs. Thus, we have

$$|Pr[W_1] - Pr[W_2]| \le Pr[F]. \tag{9}$$

We need to prove that

$$Pr[F] = ADV_{\mathbb{G},g}^{\mathrm{LCDH2}}(C), \tag{10}$$

where C is an algorithm that takes as input y_E, y_A and y_B. C will play the role of the challenger in *Game 2*. Algorithm C has a list of queries and responses, such that if A makes a query that matches one of the previous queries, C can return the previous output. At the end of the game, algorithm C will output a list with all the responses to A's queries. It is easy to see that the probability of C returning a list containing $y_E^{x_A}$ or $y_E^{x_B}$ is the same as $Pr[F]$.

Game 3. In this game we choose $\delta_A, \delta_B \xleftarrow{\$} \mathbb{Z}_q$. Adversary A will not notice the difference, since $\mathbb{PRF}$ is a pseudorandom function. Formally, there exist an algorithm B_2 such that

$$|Pr[W_2] - Pr[W_3]| = 2ADV_{\mathbb{PRF}}^{\mathrm{PRF}}(B_2). \tag{11}$$

The changes made to P_5 in *Game 1–Game 3*, transformed it into P_3. Thus, we have

$$Pr[W_3] = 1/2. \tag{12}$$

Finally, the statement is proven by combining the equalities (7)–(12).   □

---

[8] *Game 0.*
[9] *Game 1.*

Remark 4. From Theorem 6, the maximum advantage an IND-SETUP adversary can obtain in the ROM is

$$ADV_{\mathrm{DH},P_3,P_5}^{\mathrm{IND\text{-}SETUP}}(A) \leq 4\Theta ADV_{\mathbb{G},g}^{\mathrm{CDH}}(B_1) + 4\Theta ADV_{\mathrm{PRF}}^{\mathrm{PRF}}(B_2).$$

The advantage remains negligible if parameter Θ is polynomial.

4 SETUP Attacks on Legally Fair Signatures Without Keystones

To implement her attack[10], Eve works in the same environment described in Sect. 3.

As in Sect. 3, changes required by the SETUP mechanisms will further be underlined using red colored text in Fig. 6.

Description. The SETUP mechanism requires:

- a pseudorandom function $\mathbb{PRF} : K \times \mathbb{Z}_q^* \to \mathbb{Z}_q^*$, where K is the key space;
- a function $H : \mathbb{G} \to K$;
- a protocol needs to reach breakpoint ① for an attacker to recover Bob's secret key;
- a protocol needs to reach breakpoint ② for an attacker to recover both secret key.

By j_B we understand a counter incremented each time Bob runs the protocol.

The value ρ is transmitted during the protocol and is available to Eve. Hence, she can recover Alice's secret key simply by computing $\mathbb{PRF}(H(y_A^{x_E}), \rho)$ and extracting x_A from s_A by calculating $e^{-1}(\delta_A - s_A)$.

To find the value of j_B, Eve computes $\delta_{B,\ell} \leftarrow \mathbb{PRF}(H(y_B^{x_E}), \ell)$, $r_\ell = g^{\delta_{B,\ell}}$ and $\ell = \ell + 1$, until $r_\ell = r_B$. Once j_B is found, she can compute the secret key as $e^{-1}(\delta_{B,\ell} - s_B)$.

Compared to the mechanism presented in Appendix C, this SETUP attack requires only one successful protocol to recover Alice's secret key. Attacking Bob instead of Alice is less efficient, since Eve must find the current counter value.

Malicious Co-signers. If Eve is replaced by Alice, a protocol needs to reach breakpoint ①. Also, Alice may choose to infect the protocols which directly involve her, keep an internal counter, and, thus, avoid the need to use brute-force methods to obtain j_B. When replaced by Bob, a protocol needs to reach breakpoint ②.

Security Analysis. The main security results are presented in Theorems 3 and 4. The proofs are omitted given the similarity with the ones presented in Sect. 3.

---

[10] Another attack (detailed in Appendix C) naturally arises.

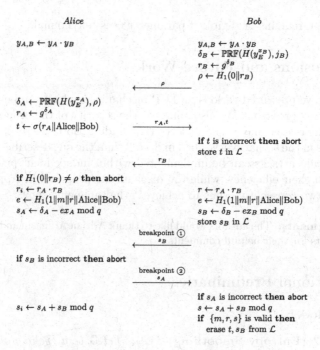

Fig. 6. The Protocol presented in Fig. 4 with a SETUP mechanism. (Color figure online)

Theorem 3. *If* DDH *is hard in* $\mathbb{G}$ *and* H *is a one-to-one function, then the protocols presented in Figs. 4 and 6 are* IND-SETUP *in the standard model. Formally, let* A *be an efficient PPT* IND-SETUP *adversary. There exist two efficient PPT algorithms* B_1, B_2 *such that*

$$ADV_{\mathrm{DH},P_4,P_6}^{\mathrm{IND\text{-}SETUP}}(A) \leq 4ADV_{\mathbb{G},g}^{\mathrm{DDH}}(B_1) + 4ADV_{\mathrm{PRF}}^{\mathrm{PRF}}(B_2).$$

Remark 5. From Theorem 3, the maximum advantage an IND-SETUP adversary can obtain in the standard model is

$$ADV_{\mathrm{DH},P_4,P_6}^{\mathrm{IND\text{-}SETUP}}(A) \leq 4\Theta ADV_{\mathbb{G},g}^{\mathrm{DDH}}(B_1) + 4\Theta ADV_{\mathrm{PRF}}^{\mathrm{PRF}}(B_2).$$

The advantage remains negligible if parameter Θ is polynomial.

Theorem 4. *If* CDH *is hard in* $\mathbb{G}$ *and* H *is a hash function, then the protocols presented in Figs. 4 and 6 are* IND-SETUP *in the ROM. Formally, let* A *be an efficient PPT* IND-SETUP *adversary. There exist three efficient PPT algorithms* B_1, B_2 *such that*

$$ADV_{\mathrm{DH},P_4,P_6}^{\mathrm{IND\text{-}SETUP}}(A) \leq 4ADV_{\mathbb{G},g}^{\mathrm{CDH}}(B_1) + 4ADV_{\mathrm{PRF}}^{\mathrm{PRF}}(B_2).$$

Remark 6. From Theorem 4, the maximum advantage an IND-SETUP adversary can obtain in the ROM is

$$ADV_{\mathrm{DH},P_4,P_6}^{\mathrm{IND\text{-}SETUP}}(A) \leq 4\Theta ADV_{\mathbb{G},g}^{\mathrm{CDH}}(C) + 4\Theta ADV_{\mathrm{PRF}}^{\mathrm{PRF}}(B_2).$$

The advantage remains negligible if parameter Θ is polynomial.

5   Conclusions and Future Work

In this paper we presented various SETUP mechanisms which can be injected in contract signing protocols. We also analyzed the security of the proposed attack scenarios. The reader may easily observe that finding Bob's secret key requires less resources in the scenario described in Sect. 3 than the one described in Sect. 4. These two main attacks can be implemented within independent protocol runs and maintain their efficiency, while the mechanisms proposed in Appendices B and C need two consecutive runs to achieve[11] the same efficiency.

Acknowledgments. The authors would like to thank Adrian Atanasiu and the anonymous reviewers for their helpful comments.

A   Additional Preliminaries

Security Models

Definition 7 (Entropy Smoothing - ES). *Let $\mathbb{G}$ be a cyclic group of order n, $\mathcal{K}$ the key space and A a PPT algorithm. Also, let $\mathcal{H} = \{h_i\}_{i \in \mathcal{K}}$ be a family of keyed hash functions, where each h_i maps $\mathbb{G}$ to $\mathbb{Z}_n^*$. We define the advantage*

$$ADV_{\mathcal{H}}^{ES}(A) = \left| Pr[A(i, h_i(z)) = 1 | i \overset{\$}{\leftarrow} \mathcal{K}, z \overset{\$}{\leftarrow} \mathbb{G}] \right.$$
$$\left. - Pr[A(i, h) = 1 | i \overset{\$}{\leftarrow} \mathcal{K}, h \overset{\$}{\leftarrow} \mathbb{Z}_n^*] \right|.$$

If $ADV_{\mathcal{H}}^{ES}(A)$ is negligible for any PPT algorithm A, we say that $\mathcal{H}$ is entropy smoothing.

Remark 7. In [8], the authors prove that the CBC-MAC, HMAC and Merkle-Damgård constructions satisfy the definition above as long as the underling primitives satisfy certain security properties.

Schnorr Signatures

ElGamal signatures [9] inspired the construction of many other DLP based signatures. We particular refer to Schnorr signatures [19] for the purpose of our current work. This family of signatures is obtained by converting interactive identification protocols into signatures[12].

We shortly describe the algorithms of the Schnorr digital signature scheme in Table 2.

---

[11] More or less.

[12] As previously described in [1,11] and implicitly used by ElGamal.

Table 2. Schnorr digital signature.

Setup(ℓ)	On input a security parameter ℓ, this algorithm selects large primes p, q such that $q \geq 2^\ell$ and $p - 1 \bmod q = 0$, as well as an element $g \in \mathbb{G}$ of order q in some multiplicative group $\mathbb{G}$ of order p, and a hash function $H_1 : \{0,1\}^* \to \{0,1\}^\ell$. The output is a set of public parameters $pp = (p, q, g, \mathbb{G}, H)$.
KeyGen(pp)	On input the public parameters pp, this algorithm chooses uniformly at random $x \xleftarrow{\$} \mathbb{Z}_q^*$ and computes $y \leftarrow g^x$. The output is the couple (sk, pk) where $sk = x$ is kept private, and $pk = y$ is made public.
Sign(pp, sk, m)	On input public parameters, a secret key sk, and a message m this algorithm selects a random $\delta \xleftarrow{\$} \mathbb{Z}_q^*$, computes $$r \leftarrow g^\delta \qquad e \leftarrow H_1(m\|r) \qquad s \leftarrow \delta - ex \bmod q$$ and outputs $\langle r, s \rangle$ as the signature of m.
Verify(pp, pk, m, σ)	On input public parameters, a public key, a message m and a signature $\sigma = \langle r, s \rangle$, this algorithm computes $e \leftarrow H_1(m, r)$ and returns True iff $g^s y^e = r$; otherwise it returns False.

B   A Supplementary SETUP Attack on Concurrent Signatures

Description. Let $H : \mathbb{G} \to \mathbb{Z}_q^*$ be a hash function. Let α be either Alice or Bob. Then, $\delta_{\alpha,0}$ represents α's secret key x_α, $r_{\alpha,0}$ represents α's public key y_α and $r_{\alpha,i} \leftarrow g^{\delta_{\alpha,i}}$. As in Sect. 3, Eve has a valid pair of keys (x_E, y_E), where y_E is stored on the victim's device.

Again, changes required by the SETUP mechanisms will further be underlined using red colored text in Fig. 7.

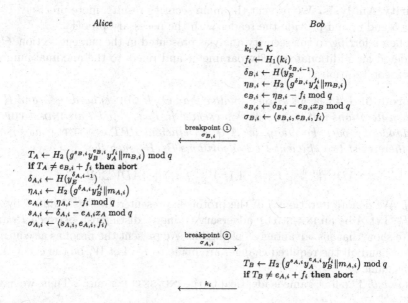

Fig. 7. Iteration i of the protocol presented in Fig. 3 with a supplementary SETUP mechanism. (Color figure online)

Eve can decide to recover Alice's secret key whenever she wants. To do that, she must first compute $\delta_{A,i} = H(r^{x_E}_{A,i-1})$. Eve recovers $r_{A,i-1}$ from an older protocol in which Alice was involved, more precisely the $i-1$ one. Thus, Eve calculates

$$g^{s_{A,i-1}} y_A^{e_{A,i-1}} \equiv g^{s_{A,i-1}+e_{A,i-1}x_A} \equiv g^{\delta_{A,i-1}} \equiv r_{A,i-1}.$$

Eve's final goal is finding x_A which can be achieved by computing $e_{A,i}^{-1}(\delta_{A,i}-s_{A,i})$. The values $e_{A,i}$ and $s_{A,i}$ are transmitted during the protocol and are public. Similarly, she can recover Bob's secret key.

The most efficient way to recover secret keys is by observing two consecutive protocol iterations that need to reach breakpoint ②.

Exceptions. An exception is iteration 1, since $\delta_{\alpha,0}$ is already known. Thus, only protocol 1 needs to reach breakpoint ②. Eve can also recover secret keys at iteration i by computing all intermediary values, $\delta_{\alpha,j}$ for $0 \le j < i$. This method is computationally costly.

Malicious Co-signers. If Eve is replaced by Alice, the most efficient way to recover secret keys is by observing two protocol iterations that need to reach breakpoint ①.

If Eve is replaced by Bob, the most efficient way to recover secret keys is by running two protocol iterations that need to reach breakpoint ②.

Security Analysis. We present the main security results, more precisely Theorems 5 and 6, and provide the reader with the necessary proofs.

When referring to the security analysis presented in the current section, Θ is considered an additional security parameter and refers to the maximal number of protocol iterations.

Theorem 5. *Let i be an integer smaller than Θ. If* DDH *is hard in $\mathbb{G}$ and H is* ES, *then iterations i of the protocols presented in Figs. 3 and 7 are* IND-SETUP *in the standard model. Formally, let A be an efficient PPT* IND-SETUP *adversary then there exist two efficient PPT algorithms B_1, B_2 such that*

$$ADV_{DH,P_3,P_7}^{IND\text{-}SETUP}(A) \le 4ADV_{\mathbb{G},g}^{DDH}(B_1) + 4ADV_{\mathcal{H}}^{ES}(B_2).$$

Proof. We denote iterations i of the protocols presented in Figs. 3 and 7 by P_3 and P_7. Let A be an IND-SETUP adversary trying to distinguish between P_3 and P_7. We show that his advantage is negligible. We present the proof as a sequence of games and all the required changes are made to P_7. Let W_i be the event that A wins game i.

Game 0. The first game is identical to the IND-SETUP game[13]. Thus, we have

$$|2Pr[W_0] - 1| = ADV_{DH,P_3,P_7}^{IND\text{-}SETUP}(A). \tag{13}$$

---

[13] see Footnote 6.

Game 1. In this game, $y_E^{\delta_{A,i-1}}$ and $y_E^{\delta_{B,i-1}}$ from *Game 0* become $g^{z_{A,i}}$ and $g^{z_{B,i}}$, where $z_{A,i}, z_{B,i} \xleftarrow{\$} \mathbb{Z}_q$. Since this is the only change between *Game 0* and *Game 1*, A will not notice the difference assuming the DDH assumption holds. Formally, this means that there exists an algorithm B_1 such that

$$|Pr[W_0] - Pr[W_1]| = 2ADV_{\mathbb{G},g}^{\text{DDH}}(B_1). \tag{14}$$

Game 2. Since H is ES then we can make the change $\delta_{A,i}, \delta_{B,i} \xleftarrow{\$} \mathbb{Z}_q$ and adversary A will not notice. Formally, this means that there exists an algorithm B_2 such that

$$|Pr[W_1] - Pr[W_2]| = 2ADV_{\mathcal{H}}^{\text{ES}}(B_2) \tag{15}$$

The changes made to P_7 in *Game 1* and *Game 2*, transformed it into P_3. Thus, we have

$$Pr[W_2] = 1/2. \tag{16}$$

Finally, the statement is proven by combining the equalities (13)–(16).   □

Remark 8. From Theorem 5, the maximum advantage an IND-SETUP adversary can obtain in the standard model is

$$ADV_{\text{DH},P_3,P_7}^{\text{IND-SETUP}}(A) \leq 4\Theta ADV_{\mathbb{G},g}^{\text{DDH}}(B_1) + 4\Theta ADV_{\mathcal{H}}^{\text{ES}}(B_2).$$

The advantage remains negligible if parameter Θ is polynomial.

Theorem 6. *Let i be an integer smaller than Θ. If CDH is hard in $\mathbb{G}$, then iterations i of the protocols presented in Figs. 3 and 7 are IND-SETUP in the ROM. Formally, let A be an efficient PPT IND-SETUP adversary then there exist an efficient PPT algorithms C such that*

$$ADV_{\text{DH},P_3,P_7}^{\text{IND-SETUP}}(A) \leq 4ADV_{\mathbb{G},g}^{\text{CDH}}(C).$$

Proof. We will use the same notations as in the proof for Theorem 5.

Game 0. The first game is identical to the IND-SETUP game[14]. Thus, we have

$$|2Pr[W_0] - 1| = ADV_{\text{DH},P_3,P_7}^{\text{IND-SETUP}}(A). \tag{17}$$

The challenger picks a random oracle $H : \mathbb{G} \to \mathbb{Z}_q^*$ at random from the set of all such functions. A can make a sequence of queries of the following type:

Hash oracle query[15]: A presents the challenger with $m \in \mathbb{G}$, who responds with $H(m)$.

Game 1. At the beginning of the game choose $z_{A,i}, z_{B,i} \xleftarrow{\$} \mathbb{Z}_q^*$. We change the challenger's way to respond to queries as follows:

Hash oracle query[16]: A presents the challenger with $m \in \mathbb{G}$. The challenger responds with:

---

[14] see Footnote 6.
[15] *Game 0.*
[16] *Game 1.*

- $z_{A,i}$, if $m = y_E^{\delta_{A,i-1}}$;
- $z_{B,i}$, if $m = y_E^{\delta_{B,i-1}}$;
- $H(m)$, otherwise.

We also make the changes $\delta_{A,i} \leftarrow z_{A,i}$ and $\delta_{B,i} \leftarrow z_{B,i}$ in P_7.

Since we have replaced the values $y_E^{\delta_{A,i-1}}$ and $y_E^{\delta_{B,i-1}}$ throughout the game, we have

$$Pr[W_0] = Pr[W_1]. \tag{18}$$

Game 2. In this game, we revert to the original hash oracle query (*i.e* the challenger responds with $H(m)$ for all m). Let F be the event that the adversary makes a query with $m \leftarrow y_E^{\delta_{A,i-1}}$ or $m \leftarrow y_E^{\delta_{B,i-1}}$. *Game 1* and *Game 2* are identical until F occurs. Thus, we have

$$|Pr[W_1] - Pr[W_2]| \leq Pr[F]. \tag{19}$$

We need to prove that

$$Pr[F] = ADV_{\mathbb{G},g}^{LCDH2}(C), \tag{20}$$

where C is an algorithm that takes as input y_E, $r_{A,i-1}$ and $r_{B,i-1}$. C will play the role of the challenger in *Game 2*. Algorithm C has a list of queries and responses, such that if A makes a query that matches one of the previous queries, C can return the previous output. At the end of the game, algorithm C will output a list with all the responses to A's queries. It is easy to see that the probability of C returning a list containing $y_E^{\delta_{A,i-1}}$ or $y_E^{\delta_{B,i-1}}$ is the same as $Pr[F]$.

The changes made to P_7 in *Game 1* and *Game 2*, transformed it into P_3. Thus, we have

$$Pr[W_2] = 1/2. \tag{21}$$

Finally, the statement is proven by combining the equalities (17)–(21).   □

Remark 9. From Theorem 6, the maximum advantage an IND-SETUP adversary can obtain in the ROM is

$$ADV_{DH,P_3,P_7}^{IND\text{-}SETUP}(A) \leq 4\Theta ADV_{\mathbb{G},g}^{CDH}(C).$$

The advantage remains negligible if parameter Θ is polynomial.

C   A Supplementary SETUP Attack on Legally Fair Signatures Without Keystones

Description. To implement an attack, Eve will work in almost the same environment as in Appendix B. Thus, we only mention the differences between the environments.

As in Sect. 3, changes required by the SETUP mechanisms are further underlined using red colored text in Fig. 8.

The most efficient way for Eve to recover secret keys is taking into account the following requirements:

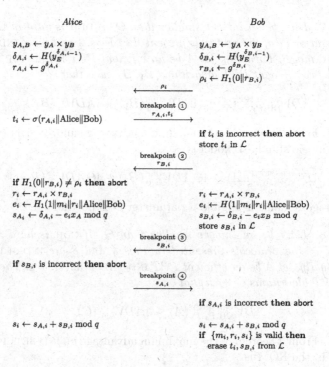

Fig. 8. Iteration i of the Protocol presented in Fig. 4 with a supplementary SETUP mechanism. (Color figure online)

1. an iteration needs to reach breakpoint ④;
2. the previous protocol iteration needs to reach breakpoint ②.

Malicious Co-signers. If Eve is replaced by Alice, the most efficient way to recover secret keys is taking into account the following requirements:

1. an iteration needs to reach breakpoint ③;
2. the previous protocol iteration needs to reach breakpoint ②.

If Eve is replaced by Bob, the most efficient way to recover secret keys is taking into account the following requirements:

1. an iteration needs to reach breakpoint ④;
2. the previous protocol iteration needs to reach breakpoint ①.

Security Analysis. The main security results are presented in Theorems 7 and 8. The proofs are omitted given their similarities with the ones constructed in Appendix B.

Theorem 7. *Let i be an integer smaller than Θ. If* DDH *is hard in* $\mathbb{G}$ *and* H *is* ES, *then iterations i of the protocols presented in Figs. 4 and 8 are* IND-SETUP *in the standard model. Formally, let A be an efficient PPT* IND-SETUP *adversary. There exist two efficient PPT algorithms B_1, B_2 such that*

$$ADV^{\text{IND-SETUP}}_{\text{DH},P_4,P_8}(A) \leq 4ADV^{\text{DDH}}_{\mathbb{G},g}(B_1) + 4ADV^{\text{ES}}_{\mathcal{H}}(B_2).$$

Remark 10. From Theorem 7, the maximum advantage an IND-SETUP adversary can obtain in the standard model is

$$ADV^{\text{IND-SETUP}}_{\text{DH},P_4,P_8}(A) \leq 4\Theta ADV^{\text{DDH}}_{\mathbb{G},g}(B_1) + 4\Theta ADV^{\text{ES}}_{\mathcal{H}}(B_2).$$

The advantage remains negligible if parameter Θ is polynomial.

Theorem 8. *Let i be an integer smaller than Θ. If* CDH *is hard in* $\mathbb{G}$, *then iterations i of the protocols presented in Figs. 4 and 8 are* IND-SETUP *in the ROM. Formally, let A be an efficient PPT* IND-SETUP *adversary. There exist an efficient PPT algorithms C such that*

$$ADV^{\text{IND-SETUP}}_{\text{DH},P_4,P_8}(A) \leq 4ADV^{\text{CDH}}_{\mathbb{G},g}(C).$$

Remark 11. From Theorem 8, the maximum advantage an IND-SETUP adversary can obtain in the ROM is

$$ADV^{\text{IND-SETUP}}_{\text{DH},P_4,P_8}(A) \leq 4\Theta ADV^{\text{CDH}}_{\mathbb{G},g}(C).$$

The advantage remains negligible if parameter Θ is polynomial.

References

1. Abdalla, M., An, J.H., Bellare, M., Namprempre, C.: From identification to signatures via the fiat-shamir transform: minimizing assumptions for security and forward-security. In: Knudsen, L.R. (ed.) EUROCRYPT 2002. LNCS, vol. 2332, pp. 418–433. Springer, Heidelberg (2002). doi:10.1007/3-540-46035-7_28
2. Asokan, N., Schunter, M., Waidner, M.: Optimistic protocols for fair exchange. In: Proceedings of the 4th ACM Conference on Computer and Communications Security (CCS 1997), pp. 7–17. ACM (1997)
3. Bellare, M., Rogaway, P.: Random oracles are practical: a paradigm for designing efficient protocols. In: Proceedings of the 1st ACM Conference on Computer and Communications Security (CCS 1993), pp. 62–73. ACM (1993)
4. Bellare, M., Rogaway, P.: Introduction to Modern Cryptography. UCSD CSE 207:207 (2005)
5. Cachin, C., Camenisch, J.: Optimistic fair secure computation. In: Bellare, M. (ed.) CRYPTO 2000. LNCS, vol. 1880, pp. 93–111. Springer, Heidelberg (2000). doi:10.1007/3-540-44598-6_6
6. Chen, L., Kudla, C., Paterson, K.G.: Concurrent signatures. In: Cachin, C., Camenisch, J.L. (eds.) EUROCRYPT 2004. LNCS, vol. 3027, pp. 287–305. Springer, Heidelberg (2004). doi:10.1007/978-3-540-24676-3_18

7. Diffie, W., Hellman, M.: New directions in cryptography. IEEE Trans. Inf. Theory **22**(6), 644–654 (2006)
8. Dodis, Y., Gennaro, R., Håstad, J., Krawczyk, H., Rabin, T.: Randomness extraction and key derivation using the CBC, Cascade and HMAC modes. In: Franklin, M. (ed.) CRYPTO 2004. LNCS, vol. 3152, pp. 494–510. Springer, Heidelberg (2004). doi:10.1007/978-3-540-28628-8_30
9. ElGamal, T.: A public key cryptosystem and a signature scheme based on discrete logarithms. IEEE Trans. Inf. Theory **31**(4), 469–472 (1985)
10. Ferradi, H., Géraud, R., Maimuţ, D., Naccache, D., Pointcheval, D.: Legally fair contract signing without keystones. In: Manulis, M., Sadeghi, A.-R., Schneider, S. (eds.) ACNS 2016. LNCS, vol. 9696, pp. 175–190. Springer, Cham (2016). doi:10.1007/978-3-319-39555-5_10
11. Fiege, U., Fiat, A., Shamir, A.: Zero knowledge proofs of identity. In: Proceedings of the 19th Annual ACM Symposium on Theory of Computing (STOC 1987), pp. 210–217. ACM (1987)
12. Garay, J., MacKenzie, P., Prabhakaran, M., Yang, K.: Resource fairness and composability of cryptographic protocols. In: Halevi, S., Rabin, T. (eds.) TCC 2006. LNCS, vol. 3876, pp. 404–428. Springer, Heidelberg (2006). doi:10.1007/11681878_21
13. Goldreich, O.: A simple protocol for signing contracts. In: Chaum, D. (ed.) Advances in Cryptology. Springer, Boston (1984). doi:10.1007/978-1-4684-4730-9_11
14. Goldwasser, S., Levin, L.: Fair computation of general functions in presence of immoral majority. In: Menezes, A.J., Vanstone, S.A. (eds.) CRYPTO 1990. LNCS, vol. 537, pp. 77–93. Springer, Heidelberg (1991). doi:10.1007/3-540-38424-3_6
15. Gordon, S.D., Hazay, C., Katz, J., Lindell, Y.: Complete fairness in secure two-party computation. J. ACM **58**(6), 1–37 (2011)
16. Lindell, A.Y.: Legally-enforceable fairness in secure two-party computation. In: Malkin, T. (ed.) CT-RSA 2008. LNCS, vol. 4964, pp. 121–137. Springer, Heidelberg (2008). doi:10.1007/978-3-540-79263-5_8
17. Micali, S.: Simple and fast optimistic protocols for fair electronic exchange. In: Proceedings of the 22nd Annual Symposium on Principles of Distributed Computing (PODC 2003), pp. 12–19. ACM (2003)
18. Pinkas, B.: Fair secure two-party computation. In: Biham, E. (ed.) EUROCRYPT 2003. LNCS, vol. 2656, pp. 87–105. Springer, Heidelberg (2003). doi:10.1007/3-540-39200-9_6
19. Schnorr, C.P.: Efficient identification and signatures for smart cards. In: Brassard, G. (ed.) CRYPTO 1989. LNCS, vol. 435, pp. 239–252. Springer, New York (1990). doi:10.1007/0-387-34805-0_22
20. Shoup, V.: Sequences of games: a tool for taming complexity in security proofs. IACR Cryptology ePrint Archive 2004, 332 (2004)
21. Simmons, G.J.: The subliminal channel and digital signatures. In: Beth, T., Cot, N., Ingemarsson, I. (eds.) EUROCRYPT 1984. LNCS, vol. 209, pp. 364–378. Springer, Heidelberg (1985). doi:10.1007/3-540-39757-4_25
22. Simmons, G.J.: Subliminal communication is easy using the DSA. In: Helleseth, T. (ed.) EUROCRYPT 1993. LNCS, vol. 765, pp. 218–232. Springer, Heidelberg (1994). doi:10.1007/3-540-48285-7_18
23. Young, A., Yung, M.: The dark side of "black-box" cryptography or: should we trust capstone? In: Koblitz, N. (ed.) CRYPTO 1996. LNCS, vol. 1109, pp. 89–103. Springer, Heidelberg (1996). doi:10.1007/3-540-68697-5_8

24. Young, A., Yung, M.: Kleptography: using cryptography against cryptography. In: Fumy, W. (ed.) EUROCRYPT 1997. LNCS, vol. 1233, pp. 62–74. Springer, Heidelberg (1997). doi:10.1007/3-540-69053-0_6

25. Young, A., Yung, M.: The prevalence of kleptographic attacks on discrete-log based cryptosystems. In: Kaliski, B.S. (ed.) CRYPTO 1997. LNCS, vol. 1294, pp. 264–276. Springer, Heidelberg (1997). doi:10.1007/BFb0052241

26. Young, A., Yung, M.: Malicious Cryptography: Exposing Cryptovirology. Wiley, New York (2004)

27. Young, A., Yung, M.: Malicious cryptography: kleptographic aspects. In: Menezes, A. (ed.) CT-RSA 2005. LNCS, vol. 3376, pp. 7–18. Springer, Heidelberg (2005). doi:10.1007/978-3-540-30574-3_2

On a Key Exchange Protocol

Mugurel Barcau[1,2], Vicenţiu Paşol[1,2], Cezar Pleşca[1,3], and Mihai Togan[1,3(✉)]

[1] certSIGN - Research and Development, Bucharest, Romania
barcau@yahoo.com, vpasol@yahoo.com, cezar.plesca@gmail.com,
mihai.togan@gmail.com
[2] Institute of Mathematics "Simion Stoilow" of the Romanian Academy,
Bucharest, Romania
[3] Military Technical Academy, Bucharest, Romania

Abstract. In this paper we investigate an instance of the generalized Diffie-Hellman key exchange protocol suggested by the equidistribution theorem. We prove its correctness and discuss the security. Experimental evidences for the theoretical results are also provided.

Keywords: Key exchange protocol · Diffie-Hellman · Rational approximations

1 Introduction

The question of key exchange is a fundamental problem in the areas of cryptography and communication security. The key exchange protocols are cryptographic primitives used to set up shared secret keys in order to enable secure communication over unreliable networks. They are the most used cryptographic tools in building secure communication protocols over the Internet (e.g. IPsec, SSH, and TLS). The first practical method for establishing such a shared secret was the Diffie-Hellman key agreement protocol, which was introduced in [3]. Much later, a generalized Diffie-Hellman algorithm was defined as a general tool for generating key exchange protocols (see [7]). The idea is very simple in essence and it can be stated as follows: assume there exist a commutative semigroup G and a set X such that G acts on X, and the action of G cannot be inverted; in the sense that if one has $x \in X$, and $g \cdot x$ (where $g \in G$), finding g is a hard task (cannot be done in polynomial time). Then, the Diffie-Hellman algorithm runs as follows: an element $x \in X$ is publicly given. Alice and Bob each choose at random secret private keys $a \in G$, $b \in G$, respectively. Alice sends $a \cdot x$ to Bob and Bob sends $b \cdot x$ to Alice. Then, Alice computes $a \cdot (b \cdot x)$, while Bob computes $b \cdot (a \cdot x)$. The associativity of the action of G on X, and the commutativity of G imply that both Alice and Bob will arrive to the same result which is set to be the common key. In this article, we investigate the key exchange protocol resulted from the action of the monoid $\mathbb{N}$ of the set of natural numbers with multiplication on the closed-open interval $[0, 1)$ of the set of real numbers, given by $(c, x) \mapsto \{cx\}$, where $\{y\}$ is the fractional part of $y \in \mathbb{R}$. Since real numbers

© Springer International Publishing AG 2017
P. Farshim and E. Simion (Eds.): SecITC 2017, LNCS 10543, pp. 187–199, 2017.
https://doi.org/10.1007/978-3-319-69284-5_13

with infinite binary representations cannot be practically used in computational algorithms, one has to use approximations of their fractional parts, so that one discovers that an approximation of this action may be the truncated product $(c, x) \mapsto \left\lfloor \frac{c \cdot x \pmod{2^n}}{2^m} \right\rfloor$ (here, $y \pmod{2^n}$ is the remainder after division of y by 2^n). A different property of this function (with x constant), more precisely its "approximately" linearity, has been used by R. Merkle in [8] to construct a key exchange protocol, which is distinct from ours.

On the other hand, a variant of the protocol constructed in this article appears in [1]. However, the authors of [1] show only experimentally that Alice and Bob get the same common key with great probability; our protocol is proven to work in all cases. We also show that the protocol is a particular instance of the generalized Diffie-Hellman key exchange protocol. Moreover, the security reduction in [1] seems faulty to us and we give two ways of showing that in fact the security of the protocol is much stronger than the security suggested in [1] (we support our conclusions also by experimental results); in particular the security reductions we construct are strong arguments for believing that the protocol is in fact a quantum-resistant protocol, which is yet another advantage over the classical Diffie-Hellman protocols, over the sought advantage of being much more efficient with respect to computational complexity.

The article is structured as follows: in Sect. 1 we present the general construction of the Diffie-Hellman protocol, as it appears in [7]. In the next section we discuss the N-monoid action described above. We shall also explain how starting with this monoid action one ends up with the truncated product function. Section 3 contains the proof of our main result, and as a consequence we describe the resulted key exchange protocol. In Sect. 4 we discuss the security of our protocol giving the necessary sizes of the parameters for practical implementation. In the next section we give experimental evidence of the fact that the truncated product function used in the algorithm cannot be inverted and also some applications of the protocol. We end the paper with a section that contains several conclusions.

2   The Generalized Diffie-Hellman Protocol

Let G be a semigroup, that is a set with an associative binary operation, denoted by "$\cdot$". In particular, we do not require that G has an identity element. The semigroup G is abelian if the operation is commutative. If S is a set, *an action of G on S* is a map

$$\phi : G \times S \longrightarrow S$$

such that $\phi(g \cdot h, s) = \phi(g, \phi(h, s)), \forall g, h \in G, s \in S$. If G is a monoid, i.e. it has identity element 1, then we shall require that $\phi(1, s) = s, \forall s \in S$. In general, we shall denote $\phi(g, s)$ by $g \cdot s$, and refer to such an action as a G-action on the set S and to S as a G-set.

We now present the key exchange protocols based on semigroup actions, as they where introduced in [7].

Protocol 1 *(Diffie-Hellman Key Exchange Protocol). Let S be a finite set, and let G be an abelian semigroup acting on S. The Diffie-Hellman Key Exchange Protocol based on the G-set S is the following protocol:*

1. *Alice and Bob publicly agree on an element $s \in S$.*
2. *Alice chooses $a \in G$ and computes $a \cdot s$. Alice's private key is a, and her public key is $a \cdot s$.*
3. *Bob chooses $b \in G$ and computes $b \cdot s$. Bob's private key is b, and his public key is $b \cdot s$.*
4. *Their common secret key is*

$$a \cdot (b \cdot s) = (a \cdot b) \cdot s = (b \cdot a) \cdot s = b \cdot (a \cdot s)$$

The above protocol is secure only if the following problem is hard:

Problem 1 (Semigroup Action Problem). Given a semigroup G acting on a set S, and elements $x \in S$ and $y \in Gx$, find $g \in G$ such that $g \cdot x = y$.

If an attacker, Eve, can find $a' \in S$ such that $a' \cdot s = a \cdot s$, then she finds the shared secret by computing: $a' \cdot (b \cdot s) = b \cdot (a' \cdot s) = b \cdot (a \cdot s)$.

Problem 2 (The Diffie-Hellman Semigroup Action Problem). Given an abelian semigroup G acting on a finite set S, and elements $x, y, z \in S$ with $y = g \cdot x$ and $z = h \cdot x$ for some $g, h \in G$, find $(g \cdot h) \cdot x \in S$.

It is clear that the security of the above protocol is equivalent to this problem. On the other hand, the only way known to attack the Diffie-Hellman Semigroup Action Problem is by solving the Semigroup Action Problem. It is unknown if these two problems are equivalent. We refer to *loc. cit.* for a detailed discussion about the generic attacks on the Semigroup Action Problem.

3 Case Study

3.1 Irrational Numbers and Equidistribution Theorem

The idea behind the Diffie-Hellman algorithm that we will study in this article is based on the well known equidistribution theorem which asserts that if x is an irrational number, then the set $\{\{n \cdot x\} \mid n \in \mathbb{N}\}$ is uniformly distributed in the interval $(0, 1)$, where $\{x\}$ stands for the fractional part of the real number x (see, [2, 13, 15].) Moreover, the monoid of natural numbers with multiplication acts on the interval $[0, 1)$ via the formula suggested by the equidistribution theorem $\mathbb{N} \times [0, 1) \mapsto [0, 1), (n, x) \mapsto \{n \cdot x\}$:
It is very easy to check that indeed,

$$\{m \cdot \{n \cdot x\}\} = \{mn \cdot x\} = \{n \cdot \{m \cdot x\}\}.$$

There are two issues to be resolved concerning this example. The first one is how do we represent the irrational numbers in order to do practical computation.

And the second issue (which is obviously related to the first one) is how certain we are that the corresponding Diffie-Hellman algorithm is secure. There are two alternatives to represent a real number: by symbols or by approximation.

The first alternative implies that the number is considered as a solution of certain algebraic/differential equations, e.g. we represent $\sqrt{2}$ as the unique positive solution of the equation $x^2 - 2 = 0$. This type of representation is not suitable for our purposes since the representation of $\{n \cdot x\}$ would reveal n, or, even worse, would be impossible even to compute $\{n \cdot x\}$ for large n.

On the other hand the alternative of approximating a real number (by some finite expression) seems to be also doomed since that representation would actually represent a rational number (in any of the natural known representations, i.e. by digits in some base, by continued fraction, etc.). But then, we will loose the nice property of uniform distribution of the numbers $\{n \cdot x\}$ when n varies, which should be important for proving the randomness of the algorithm.

We will choose the second alternative and see that in fact, the randomness property is not entirely lost, but rather propagates well enough to prove the security of the algorithm.

3.2   Base 2 Approximation of Subunitary Numbers and Merkle's Approximately Linear Hash Function

We choose base 2 approximation because it is the most suited for computer manipulations. Let n be a natural number (to be setup later) and consider an irrational number $x \in (0, 1)$. We write $\bar{x} = \bar{x}_n \in \{0, \ldots, 2^n - 1\}$ its base 2, n-digit expansion i.e. $\bar{x} = \lfloor 2^n x \rfloor$, where $\lfloor y \rfloor$ stands for the integer part of a real number y. In general, for any positive real number, we write $\bar{x} = \bar{x}_n := \lfloor 2^n x \rfloor \pmod{2^n}$. We omit the index n in the notation of $\bar{x}$ if it is obvious from the context. If a is a k-digit number ($k \le n$), then the $(n - k)$-bit expansion of $\{ax\}$ will be almost $\lfloor (a\bar{x} \pmod{2^n})/2^k \rfloor$, where by $a\bar{x} \pmod{2^n}$ we mean the remainder from division of $a\bar{x}$ by 2^n. One should notice that the function $a \mapsto \lfloor (a\bar{x} \pmod{2^n})/2^k \rfloor$ is the approximately linear function $AL(a, x)$ in [8]. It will become clear in the next sections how good is the last approximation (it can differ by 1 at most). Observe that if we publicly publish x of size n, then $\overline{\{ax\}}_n$ determines a if small enough, thus we cannot use this function for a key exchange protocol. However, if we cut the last k digits, where k is the size of a, the function becomes not invertible, as we shall see in the section dedicated to the security of the protocol. We have now in our hands, indeed the tools (commutative semigroup action and hardness of semigroup action problem) in order to produce the Diffie-Hellman protocol.

4   Key Exchange Protocol

For any $a \in [0, 2^k - 1]$, $x \in [0, 2^n - 1]$ and positive integer $m \le n$ we define the function

$$\phi_{(k,n,m)}(a, x) := \left\lfloor 2^m \left\{ \frac{ax}{2^n} \right\} \right\rfloor = \left\lfloor \frac{ax \pmod{2^n}}{2^{n-m}} \right\rfloor$$

(here, as before, by $ax \pmod{2^n}$ we mean the remainder from division of ax by 2^n).

Theorem 2. *Let k, n, m, l be positive integers such that $m \geq k + l$. For any $a, b \in [0, 2^k - 1]$, $x \in [0, 2^n - 1]$ there exists $\delta \in \{-1, 0, 1\}$ such that*

$$\phi_{(k,m,l)}(a, \phi_{(k,n,m)}(b, x)) \equiv \phi_{(k,m,l)}(b, \phi_{(k,n,m)}(a, x)) + \delta \pmod{2^l}$$

Proof. We make the following notations:

$$x_A := \phi_{(k,n,m)}(a, x) \in [0, 2^m - 1], x_B := \phi_{(k,n,m)}(b, x) \in [0, 2^m - 1]$$

Since $\dfrac{ax}{2^n} = \left\lfloor \dfrac{ax}{2^n} \right\rfloor + \left\{ \dfrac{ax}{2^n} \right\}$ we get:

$$\frac{ax}{2^{n-m}} = 2^m \left\lfloor \frac{ax}{2^n} \right\rfloor + 2^m \left\{ \frac{ax}{2^n} \right\} = 2^m \left\lfloor \frac{ax}{2^n} \right\rfloor + x_A + \left\{ 2^m \left\{ \frac{ax}{2^n} \right\} \right\},$$

which yields the inequalities:

$$0 \leq \frac{ax}{2^n} - \left\lfloor \frac{ax}{2^n} \right\rfloor - \frac{x_A}{2^m} < \frac{1}{2^m}$$

Now, denote by $x_{AB} := \phi_{(k,m,l)}(a, x_B)$, and by $x_{BA} := \phi_{(k,m,l)}(b, x_A)$, then we have:

$$0 \leq \frac{abx}{2^n} - b \left\lfloor \frac{ax}{2^n} \right\rfloor - \frac{bx_A}{2^m} < \frac{b}{2^m}$$

$$0 \leq \frac{abx}{2^n} - b \left\lfloor \frac{ax}{2^n} \right\rfloor - \left\lfloor \frac{bx_A}{2^m} \right\rfloor - \left\{ \frac{bx_A}{2^m} \right\} < \frac{b}{2^m}$$

$$0 \leq \frac{abx}{2^{n-l}} - 2^l \left(b \left\lfloor \frac{ax}{2^n} \right\rfloor + \left\lfloor \frac{bx_A}{2^m} \right\rfloor \right) - 2^l \left\{ \frac{bx_A}{2^m} \right\} < \frac{b}{2^{m-l}}$$

$$0 \leq \frac{abx}{2^{n-l}} - 2^l \left(b \left\lfloor \frac{ax}{2^n} \right\rfloor + \left\lfloor \frac{bx_A}{2^m} \right\rfloor \right) - \left\lfloor 2^l \left\{ \frac{bx_A}{2^m} \right\} \right\rfloor - \left\{ 2^l \left\{ \frac{bx_A}{2^m} \right\} \right\} < \frac{b}{2^{m-l}}$$

$$0 \leq \frac{abx}{2^{n-l}} - 2^l \left(b \left\lfloor \frac{ax}{2^n} \right\rfloor + \left\lfloor \frac{bx_A}{2^m} \right\rfloor \right) - x_{BA} < \frac{b}{2^{m-l}} + \left\{ 2^l \left\{ \frac{bx_A}{2^m} \right\} \right\}$$

Since $m \geq k + l$, we deduce $0 \leq \dfrac{b}{2^{m-l}} + \left\{ 2^l \left\{ \dfrac{bx_A}{2^m} \right\} \right\} < 2$, so that:

$$\left\lfloor \frac{abx}{2^{n-l}} \right\rfloor = 2^l \alpha_{BA} + x_{BA} + \epsilon_{BA} \tag{1}$$

for some integer α_{BA} and $\epsilon_{BA} \in \{0, 1\}$. Similarly we have:

$$\left\lfloor \frac{abx}{2^{n-l}} \right\rfloor = 2^l \alpha_{AB} + x_{AB} + \epsilon_{AB} \tag{2}$$

for some integer α_{AB} and $\epsilon_{AB} \in \{0, 1\}$. From (1) and (2) we get the congruence:

$$x_{AB} \equiv x_{BA} + \delta \pmod{2^l}$$

where $\delta \in \{-1, 0, 1\}$.

Notice that if $l \geq 2$ the last congruence gives $x_{AB} \equiv x_{BA} + \delta \pmod{2^2}$, which means that the last two digits in the binary decompositions of x_{AB} and x_{BA} determine δ. This simple but important observation is included in the following key exchange protocol.

1. **Public key:** Choose n, k, m, l such that $n \geq m \geq k + l$. Pick a random *good* number $x \in [0, 2^n - 1]$. The public key is (n, k, m, l, x).
2. **Secret choices:** Alice picks a random number $a \in [0, 2^k - 1]$ and Bob picks a random number $b \in [0, 2^k - 1]$.
3. **Exchange:** Alice computes $x_A := \left\lfloor \dfrac{ax \pmod{2^n}}{2^{n-m}} \right\rfloor$ and Bob computes $x_B :=$ $\left\lfloor \dfrac{bx \pmod{2^n}}{2^{n-m}} \right\rfloor$. Alice sends x_A to Bob and Bob sends x_B to Alice.
4. **Verify key:** Alice computes $x_{AB} := \left\lfloor \dfrac{ax_B \pmod{2^m}}{2^{m-l}} \right\rfloor$, then she sets (k_A, v_A) to be the most significant $l - 2$ digits, respectively the least significant 2 digits of x_{AB}. Similarly, Bob computes (k_B, v_B). Both publicly publish v_A and v_B, respectively.
5. **Common key:** If $(v_A, v_B) = (00, 11)$ then the common key is $K := k_A = (k_B + 1) \pmod{2^{l-2}}$. If $(v_A, v_B) = (11, 00)$ then the common key is $K := (k_A + 1) \pmod{2^{l-2}} = k_B$. Otherwise, the common key is $K := k_A = k_B$.

By a *good* number, we mean an odd number whose distribution of $0'$s and $1'$s in its binary expansion is random.

Notice that the value $n - m$ must be large enough to be resistant to brute force attacks, thus from the security perspective, the choice of n, m, and l has to be such that $n \geq m + k \geq 2k + l$.

4.1 Security

As explained in Sect. 2, the security of our protocol is based on the hardness of inverting the function $a \mapsto \left\lfloor \dfrac{ax \pmod{2^n}}{2^{n-m}} \right\rfloor$, where x is a known (good) n-digit number. The authors in [1] suggest that the hardness of this problem can be reduced to an instance of SAT by explicitly writing down the equations for the digits of a and the carry-overs, and comparing those equations with the equations used in [5] to instantiate SAT from FACT which is believed to be classically hard. However, Merkle's 3SAT reduction, see [8], suggests that the hardness of the problem is in fact based on an NP complete problem, which indicates that the problem might be also quantum secure. Our personal take is towards Merkle's point of view. Moreover, the authors in [1] seem to overlook some facts about the shape of the equations in their comparison with FACT. To give a stronger argument why we are inclined towards Merkle's point of view, we argue that the problem is more related to CVP (closest vector problem in a lattice) than to FACT. Let $y = \left\lfloor \dfrac{ax \pmod{2^n}}{2^{n-m}} \right\rfloor$, then there exist $q \in \mathbb{Z}$ and

$r \in [0, 2^{n-m} - 1]$ such that

$$\frac{ax}{2^n} - q = \frac{y}{2^m} + \frac{r}{2^n}.$$

Thus, one has to find the closest vector to $\frac{y}{2^m}$ in the "lattice" $\tilde{\Lambda}_x := \mathbb{Z} \cdot \frac{x}{2^n} + \mathbb{Z}$. The fact that the point $\frac{y}{2^m} + \frac{r}{2^n}$ is (probabilistically) the closest vector in $\tilde{\Lambda}_x$ for sufficiently random public key x is implied by the fact that the function $a \mapsto y$ is probabilistically injective as shown bellow.

The "lattice" $\tilde{\Lambda}_x$ is "an approximation" of the lattice $\Lambda_\alpha := \mathbb{Z}\alpha + \mathbb{Z}$, for α an irrational number. Notice that the later lattice corresponds to a noncommutative elliptic curve $E_\alpha := \mathbb{R}/\Lambda_\alpha$ (see [14]) and the action of the monoid of natural numbers acts as usual multiplication (successive additions) on E_α. One may argue now that in CVP the dimension of the lattice is important for the hardness of the problem. Note that in fact, by approximating the lattice and taking only the most important bits in this approximation, the "dimension" of the "lattice" can be considered to be $2k$ (this is the number of free variables over $\mathbb{F}_2$), which is in agreement with the usual setup of CVP.

4.2 Hashing Perspective

Another perspective over the security of the exchange protocol could be seen by considering Alice's computation x_A as a multiplicative hash function of key $a \in [0, 2^k - 1]$, where x, n and m are constant parameters:

$$x_A := \left\lfloor \frac{ax \;(\mathrm{mod}\; 2^n)}{2^{n-m}} \right\rfloor$$

A simplified version of the general multiplicative hash function was proposed by Dietzfelbinger et al. [9] and consists in obtaining a hash value of size m for an integer key $a \in [0, 2^n - 1]$, using the previous formula: $h(a) = x_A$. The authors show that if x is a randomly chosen odd integer in $[0, 2^n - 1]$, then the collision probability of two different keys is almost $2/2^m$, which is a factor of two larger than what one could expect with a random function from $2^n \to 2^m$.

Now, let's consider $a \in [0, 2^k - 1]$ as a fixed value. Given the above mentioned collision probability, it means that if we randomly choose $a' \in [0, 2^n - 1]$, the probability to collude with a (i.e. $h(a) = h(a')$) is almost $2/2^m$. One can easily notice that, knowing the approximately linear behavior of the hash function, if we restrain a' to the range $[0, 2^k - 1]$, the collision probability remains $2/2^m$.

Let X be a random variable that counts the number of collisions for function h in the range $[0, 2^k - 1]$. Using a technique similar to the analysis of the birthday paradox [10], the expected number of collisions is limited to:

$$E[X] \le \binom{2^k}{2} \frac{2}{2^m}$$

Now applying Markov's inequality, we have: $\mathrm{Prob}(X \ge 1) \le E[X]$. This implies that one can approximate the probability of no collision in all 2^k keys:

$$\mathrm{Prob}(\text{no collision}) = \mathrm{Prob}(X = 0) = 1 - \mathrm{Prob}(X \ge 1) \ge 1 - E[X]$$

$$\implies \text{Prob(no collision)} \geq 1 - \binom{2^k}{2} \frac{2}{2^m} = 1 - \frac{2^k(2^k - 1)}{2} \cdot \frac{2}{2^m} > 1 - 2^{m-2k}$$

Therefore, taking for instance $m \geq 3k$, the hash function h becomes probabilistically injective, which implies that no further reductions can be made by an attacker on a brute force verification over the range $[0, 2^k - 1]$. Finally, we can conclude that the security parameter of the presented protocol is k.

However, the experiments show that in fact the conclusions of this subsection are in fact valid for smaller values, i.e. $m = 2k + 2$(see the bellow experimental results).

5   Experiments

5.1   Key Distribution

An important issue in the application of this protocol in practice is related to the randomness of the common secret key. In order to test this property, we conduct an experiment using a library for arbitrary-precision integer math, namely BIGNUM library that comes with OpenSSL [11].

As we want to obtain shared secret keys of length 128 bits, we have chosen the following values for the scheme parameters: $k = 128$, $l = k + 2 = 130$, $m = 2k + 2 = 258$ and $n = 3k + 2 = 386$. We pick three random numbers (using three independent generators): $x \in [0, 2^n - 1]$, $a \in [0, 2^k - 1]$ (Alice choice) and $b \in [0, 2^k - 1]$ (Bob choice). Then, using the protocol described in Sect. 4, we compute the shared secret key $s = k_A = k_B$. For such an execution, we also count if the protocol needs an additional step at the end to adjust the keys of Alice or Bob. We called this a key adjustment.

As the number of possible common keys is very large, we divide the range $[0, 2^{128} - 1]$ into 128 equal bins. We repeat the previous execution a number of $N = 10^8$ times, and for each execution, we place the secret key into its corresponding bin. The percentage of key adjustment cases is about $6, 6\%$. Finally, we count the number of keys belonging to each bin, and normalize these frequencies. As expected, the keys distribution is almost uniform as shown in Fig. 1. The mean square error between our distribution and the ideal one is approximatively $7 * 10^{-9}$.

In order to verify the theoretical results presented in Sect. 4.2, we conduct another experiment, to obtain the collision probability for the hash function $h(a) = x_A : [0, 2^k - 1] \rightarrow [0, 2^m - 1]$. We vary the length of the shared key k in the range $[10 - 20]$; the other parameters are computed as previously: $l = k + 2$, $m = 2k + 2$ and $n = 3k + 2$.

For each k, using three independent random generators for $x \in [0, 2^n - 1]$, $a \in [0, 2^k - 1]$ and $b \in [0, 2^k - 1]$, we compute x_A and x_B. Theoretically, we'll have a collision (i.e. $x_A = x_B$) with a probability inferior to $1/2^{m-1}$. We execute this experience a number of $N = 2^{m-1} \cdot 1000$ times, count the number of collisions and convert this number into a probability.

The results obtained are illustrated in Fig. 2, using a logarithmic scale for collision probability. One can notice that the collision probability is a factor of two smaller than the theoretical limit $1/2^{m-1}$. In other words, this empirical probability (i.e. $1/2^m$) is equivalent to what one could expect with a random function from $2^k \to 2^m$. Building on this result, we can assume that the function x_A is statistically injective, which confirms the theoretical results from Sect. 4.2.

5.2 Rough Distributions

We will show in what follows why the probability of key adjustment agrees with the empirical data we produced.

As in the previous sections, we have the following ($m = n - k$):

$$0 \le \frac{abx}{2^{n-l}} - 2^l \left(b \left\lfloor \frac{ax}{2^n} \right\rfloor + \left\lfloor \frac{b \cdot x_A}{2^{n-k}} \right\rfloor \right) - x_{BA} = \frac{b \cdot r_A}{2^{n-l}} + \left\{ 2^l \left\{ \frac{b \cdot x_A}{2^{n-k}} \right\} \right\}.$$

Taking the integer parts, we get:

$$\left\lfloor \frac{abx}{2^{n-l}} \right\rfloor - 2^l \left(b \left\lfloor \frac{ax}{2^n} \right\rfloor + \left\lfloor \frac{b \cdot x_A}{2^{n-k}} \right\rfloor \right) = x_{BA} + \frac{b \cdot r_A}{2^{n-l}} + \left\{ 2^l \left\{ \frac{b \cdot x_A}{2^{n-k}} \right\} \right\}.$$

Modulo 2^l the left hand side is the same for Alice and Bob, therefore for a key adjustment to take place, the right hand side has to change. Since

$$\left\{ 2^l \left\{ \frac{b \cdot x_A}{2^{n-k}} \right\} \right\} = \left\{ \frac{abx \pmod{2^n}}{2^{n-l}} - \frac{b \cdot r_A}{2^{n-l}} \right\}.$$

Thus in order to have a change on the right hand side either $\left\{ \frac{abx \pmod{2^n}}{2^{n-l}} - \frac{b \cdot r_A}{2^{n-l}} \right\} + \frac{b \cdot r_A}{2^{n-l}} > 1$ and $\left\{ \frac{abx \pmod{2^n}}{2^{n-l}} - \frac{a \cdot r_B}{2^{n-l}} \right\} + \frac{a \cdot r_B}{2^{n-l}} < 1$ or the other way around.

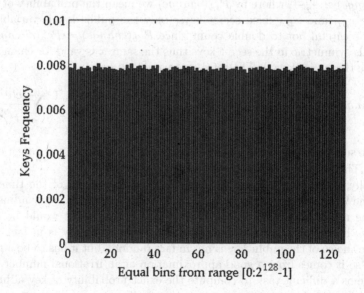

Fig. 1. Shared secret keys distribution

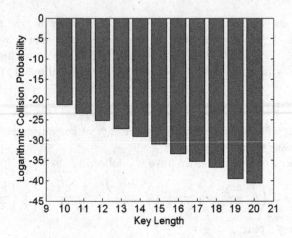

Fig. 2. Collision probability of hash function

Notice that $\{u - v\} + v > 1 \Leftrightarrow \{u\} < v$ for any real u and $0 < v < 1$. We get:

$$\frac{a \cdot r_B}{2^{n-l}} \leq \left\{ \frac{abx \pmod{2^n}}{2^{n-l}} \right\} < \frac{b \cdot r_A}{2^{n-l}},$$

or the other way around. That means that $abx \pmod{2^n}$ sits in a short interval. As one can see, the length of this interval depends upon the choice of a and b. However, on average (over x good), the length of the interval is around $\frac{|b-a|}{2^{n-l}}$. Using the uniform distribution of $abx \pmod{2^n}$ when x varies, we get that $P_{a,b}(change) \simeq \frac{|b-a|}{2^k}$, where by $P_{a,b}(change)$ we mean the probability of having different x_{AB} and x_{BA} fixed secret keys (only the public key is variable). We should be careful not to double count since $P_{a,b}(change) = P_{b,a}(change)$ (the interval is symmetric in the secret keys thus the same x is valid for change when we swipe the secret keys.). The total average would be:

$$P(change) = \frac{1}{2^{2k}} \frac{1}{2} \sum_{a,b} P_{a,b}(change) \simeq \frac{1}{2^{3k+1}} \sum_{a,b} |b - a| = \frac{1}{2^{3k}} \sum_{c} \frac{c(c+1)}{2},$$

The last sum equals $\frac{1}{2^{3k}} \frac{M(M+1)(M+2)}{3}$ where $M = 2^k - 1$ is the range of summation. This gives a rough approximation of $P(change) = \frac{1}{3}$.

Finally, notice that the key adjustment occurs only $1/4$ of the times since out of the 8 possibilities for (v_{AB}, v_{BA}), only 2 of them produce key adjustment. Thus the rough probability (on average) of key adjustment would be around $1/12$. We have to be notice that in practice, this probability is in fact smaller since the choice of the public key is not in fact random, but it has to be a random number so it comes from a good approximation of an irrational number.

It seems a difficult task to compute the exact probability of key adjustment and we leave this computation as an open question.

5.3 Comparison with DLP-Based Diffie-Hellman Protocol

An important issue is also the computation effort required to run the protocol phases. This could have severe implications, especially in the case of resources with low power consumption requirements (like IoT devices) or in the cases where the millions of secret keys must be exchanged during a short time period. The key exchange protocol presented above is much less computing intensive than Diffie-Hellman protocol based on classical discrete logarithm problem over Z_p. To achieve a common secret, each party has to make only two multiply operation on integers, the first one to generate the information X_A that has to be exchanged, and the second one to compute the verify key X_{AB}. Truncations are also used to discard the first and the last k-bits of the result, but these are almost free in terms of computing costs. Besides this, the initial setup of the protocol required to generate the parameters and the public/secret keys are also free in terms of computational effort.

For comparison purpose, we conducted few experiments intended to estimate the computation efforts required by our key-exchange protocol by comparison with DLP-based DH and ECDH protocols. We used an OpenSSL BIGNUM library based implementation for our protocol and the reference implementations for DH and ECDH variants included also in the OpenSSL package.

To have a relevant comparison, we have used similar configurations for the security-bits level (it is established by k value in our protocol), and identical length for the computed secret keys. For an equivalent security parameter k and a given length L_{SK} for the outputted secret key, we instantiated the protocol with a setup based on the following parameters: $l = L_{SK} + 2$, $m = k + L_{SK} + 2$ and $n = 2k + L_{SK} + 2$. To estimate the required computation effort, we measured the costs in terms of the processing time of the steps that compute the common secret key. The initial phases of the protocol required for the generation of the parameters, public and private keys have not been taken into account.

In the case of DH protocol over $\mathbb{F}_p$ and using a modulus p of 2048-bits length (this leads to a security level of $k = 112$-bits [12]), our protocol was 4500 times faster. In the case of DH with a modulus of 3072-bits length (security level of 128-bits), our protocol was 10000 times faster. Even DH on elliptic curves is less expansive that DH on $\mathbb{F}_p$, our key-exchange variant is about 1000 times faster than ECDH on prime fields $\mathbb{F}_p$, and respectively about 2000 times faster than ECDH on binary fields $\mathbb{F}_{2^m}$. Our experiments were conducted on a machine based on Intel(R) Xeon(R) E5-1620 at 3.60GHz CPU. In the case of much slower computing resources (such as IoT enabling devices), we expect that the mentioned speed-up rates will be much higher.

6 Conclusions

The observations in Sect. 4.1 might suggest new types of key exchange protocols by considering rational approximations for other geometric meaningful encryption algorithms. The advantage of rational approximations would be two folded: computational complexity relaxation of the usual encryption algorithms and additional security via a CVP-type argument. The theoretical and experimental results in this paper make our protocol valuable for the cases where slow computing resources are available as mentioned above. Further work will concentrate on improving the theoretical security bounds as suggested by the experimental results and on the practical implementation of the protocol.

After the write-up of the paper we learned that there exists an attack of the key exchange protocol presented in this paper using an embedding of the security problem of our protocol into a two dimensional Closest Vector Problem (see for instance [16]). More precisely, one has a precise description of a two dimensional lattice and the constraints in our protocol ask for finding a lattice vector of bounded distance. Moreover, the exact conditions imposed on the parameters imply that one has a unique solution, thus one can test the enumerated lattice vectors output by the Bounded Distance Decoding algorithm (see [6]) and find a solution of our security problem. Unfortunately, one cannot modify the parameters in our protocol so the attack becomes unfeasible.

Acknowledgments. This research was partially supported by the Romanian National Authority for Scientific Research (CNCS-UEFISCDI) under the project PN-III-P2-2.1-PTE-2016-0191.

References

1. Azhari, A., Bouftass, S.: On a new fast public key cryptosystem. https://eprint.iacr.org/2014/946.pdf
2. Bohl, P.: Über ein in der Theorie der säkutaren Störungen vorkommendes Problem. J. Reine Angew. Math. **135**, 189–283 (1909)
3. Diffie, W., Hellman, M.: New directions in cryptography. IEEE Trans. Inf. Theory **22**(6), 644–654 (1976)
4. Gerold Grünauer Proposal of a new efficient public key system for encryption and digital signatures. https://eprint.iacr.org/2007/445.pdf
5. Horie, S., Watanabe, O.: Hard instance generation for SAT. In: Leong, H.W., Imai, H., Jain, S. (eds.) ISAAC 1997. LNCS, vol. 1350, pp. 22–31. Springer, Heidelberg (1997). doi:10.1007/3-540-63890-3_4
6. Liu, Y.-K., Lyubashevsky, V., Micciancio, D.: On bounded distance decoding for general lattices. In: Díaz, J., Jansen, K., Rolim, J.D.P., Zwick, U. (eds.) APPROX/RANDOM -2006. LNCS, vol. 4110, pp. 450–461. Springer, Heidelberg (2006). doi:10.1007/11830924_41
7. Maze, G., Monico, C., Rosenthal, J.: Public key cryptography based on semigroup actions. Adv. Math. Commun. **1**(4), 489–507 (2007)
8. Merkle, R.C.: Public key distribution using approximately linear functions. http://www.merkle.com/papers/approxLinearPK.html
9. Dietzfelbinger, M., Hagerup, T., Katajainen, J., Penttonen, M.: A reliable randomized algorithm for the closest-pair problem. J. Algorithms **25**(1), 19–51 (1997)
10. Cormen, T.H., Leiserson, C.E., Rivest, R.L., Stein, C.: Introduction to Algorithms, 3rd edn. MIT Press, Cambridge (2009)
11. Serpette, B., Vuillemin, J., Hervé, J.-C.: BigNum: a portable and efficient package for arbitrary-precision arithmetic. Digital, Paris Research Laboratory (1989)

12. NIST: Recommendation for Key Management, NIST Special Publication 800–57 Part 1 Revision 4 2016
13. Sierpinski, W.: Sur la valeur asymptotique d'une certaine somme. Bull Intl. Acad. Polonmaise des Sci. et des Lettres (Cracovie) series A, 9–11 (1910)
14. Soibelman, Y.: Quantum tori mirror symmetry and deformation theory. Lett. Math. Phys. **56**(2), 99–125 (2001)
15. Weyl, H.: Über die Gibbs'sche Erscheinung und verwandte Konvergenzphänomene. Rendiconti del Circolo Matematico di Palermo, pp. 377–407 (1910)
16. Zhang, Y.: A practical attack to Bouftass's crypto system. https://arxiv.org/abs/1605.00987v1

Author Index

Printed in the United States
by Baker & Taylor

Printed in the United States
By Bookmasters